T0073045

Introduction to Medical Software

Providing a concise and accessible overview of the design, implementation, and management of medical software, this textbook will equip students with a solid understanding of critical considerations for both standalone medical software (software as a medical device/SaMD) and software that is integrated into hardware devices. It includes: practical discussion of key regulatory documents and industry standards, and how these translate into concrete considerations for medical software design; detailed coverage of the medical software lifecycle process; an accessible introduction to quality and risk management systems in the context of medical software; succinct coverage of essential topics in data science, machine learning, statistics, cybersecurity, software engineering, and healthcare, bringing readers up to speed; six cautionary real-world case studies which illustrate the dangers of improper or careless software processes. Accompanied by online resources for instructors, this is the ideal introduction for undergraduate students in biomedical engineering, electrical engineering and computer science, junior software engineers, and digital health entrepreneurs.

Xenophon Papademetris is Professor of Radiology and Biomedical Imaging, and Biomedical Engineering, at Yale University. His research focuses on biomedical image analysis and software development. He has over twenty years of experience in medical software design.

Ayesha N. Quraishi is an MD serving as the Chief of Otolaryngology at the Cambridge Health Alliance; Faculty at Harvard Medical School; former Clinical Director of the Yale Center for Biomedical Innovation and Technology; member of the MIT H@cking Medicine Initiative and Lecturer in the Practice of Management at Yale School of Management. She holds an MBA from the MIT Sloan School of Management with an emphasis on global leadership and innovation.

Gregory P. Licholai teaches at Yale School of Management and is Co-Director of the Center for Digital Health. He also co-teaches Innovating Health Care at Harvard Business School. He is Chief Medical and Innovation Officer at ICON plc, a leading pharmaceutical service, healthcare data, and contract research provider. He serves on multiple company and non-profit boards including the Digital Medicine Society (DiMe) and is an advisor on the Clinical Trials Transformation Initiative (CTTI), a public–private partnership co-founded by Duke University and the Food and Drug Administration (FDA). He received his MD from Yale School of Medicine and MBA from Harvard Business School. He writes about innovation in healthcare for *Forbes*.

Cambridge Texts in Biomedical Engineering

Series Editors
W. Mark Saltzman, *Yale University*
Shu Chien, *University of California, San Diego*

Series Advisors
Jerry Collins, *Alabama A & M University*
Robert Malkin, *Duke University*
Kathy Ferrara, *University of California, Davis*
Nicholas Peppas, *University of Texas, Austin*
Roger Kamm, *Massachusetts Institute of Technology*
Masaaki Sato, *Tohoku University, Japan*
Christine Schmidt, *University of Florida*
George Truskey, *Duke University*
Douglas Lauffenburger, *Massachusetts Institute of Technology*

Cambridge Texts in Biomedical Engineering provide a forum for high-quality textbooks targeted at undergraduate and graduate courses in biomedical engineering. They cover a broad range of biomedical engineering topics from introductory texts to advanced topics, including biomechanics, physiology, biomedical instrumentation, imaging, signals and systems, cell engineering, and bioinformatics, as well as other relevant subjects, with a blending of theory and practice. While aiming primarily at biomedical engineering students, this series is also suitable for courses in broader disciplines in engineering, the life sciences and medicine.

Introduction to Medical Software

Foundations for Digital Health, Devices, and Diagnostics

Xenophon Papademetris
Yale University, Connecticut

Ayesha N. Quraishi
Harvard Medical School and Yale University, Connecticut

Gregory P. Licholai
Yale University, Connecticut

CAMBRIDGE
UNIVERSITY PRESS

CAMBRIDGE
UNIVERSITY PRESS

University Printing House, Cambridge CB2 8BS, United Kingdom

One Liberty Plaza, 20th Floor, New York, NY 10006, USA

477 Williamstown Road, Port Melbourne, VIC 3207, Australia

314–321, 3rd Floor, Plot 3, Splendor Forum, Jasola District Centre,
New Delhi – 110025, India

103 Penang Road, #05–06/07, Visioncrest Commercial, Singapore 238467

Cambridge University Press is part of the University of Cambridge.

It furthers the University's mission by disseminating knowledge in the pursuit of
education, learning, and research at the highest international levels of excellence.

www.cambridge.org
Information on this title: www.cambridge.org/highereducation/isbn/9781316514993
DOI: 10.1017/9781009091725

© Xenophon Papademetris, Ayesha N. Quraishi, and Gregory P. Licholai 2022

First published 2022

A catalogue record for this publication is available from the British Library.

Library of Congress Cataloging-in-Publication Data
Names: Papademetris, Xenophon, 1971– author. | Quraishi, Ayesha N., 1975–
    author. | Licholai, Gregory P., 1964– author.
Title: Introduction to medical software : foundations for digital health, devices, and
    diagnostics / Xenophon Papademetris, Ayesha N. Quraishi, Gregory P. Licholai.
Other titles: Cambridge texts in biomedical engineering
Description: Cambridge, United Kingdom ; New York, NY : Cambridge University
    Press, 2022. | Series: Cambridge texts in biomedical engineering |
    Includes bibliographical references and index.
Identifiers: LCCN 2022000137 | ISBN 9781316514993 (hardback)
Subjects: MESH: Software | Medical Informatics | Artificial Intelligence |
    BISAC: TECHNOLOGY & ENGINEERING / Engineering (General)
Classification: LCC R855.3 | NLM W 26.55.S6 | DDC 610.285–dc23/eng/202202046
LC record available at https://lccn.loc.gov/2022000137

ISBN 978-1-316-51499-3 Hardback

Additional resources for this publication at www.cambridge.org/medicalsoftware

Contents

Preface

The goal of this book is to provide in one brief and accessible volume a survey of the critical material involved in the design, implementation, and management of medical software for both standalone software ("software as a medical device – SaMD") and software that is part of a physical medical device. One will find more detailed treatments of many of the topics covered in this book in specialized books that focus on some of the topics we cover (e.g. software engineering, systems engineering, probability theory, machine learning). Depth was not our goal; this book is explicitly designed to provide a broad survey.

Our hope is to familiarize the reader with the span of topics he or she may need in entering this field and to provide pointers to more specialized publications as this becomes necessary. For example, most computer scientists have very limited exposure to statistical decision theory, and we think that even the cursory coverage in this book will at least enable them to understand "what they do not know" and seek help as opposed to being ignorant of this entire field and attempting to reinvent the wheel in an amateurish manner!

An emerging challenge in medical software is the increasing use of big data and artificial intelligence/machine learning (AI/ML) techniques. This places an even greater stress on proper software design and management. Given that these are "black box" methods, in which the human understanding of what actually is going on is limited, a proper software quality process will be even more critical in creating reliable software tools. We introduce this topic in Section 1.3. In that section we also provide pointers to the other sections of the book in which we discuss issues related to the use of AI/ML methods.

This is an introductory book. One can and should follow the material here with further study, using both original regulatory materials, industry standards,[1] and more advanced books.[2] Our goal can be summarized by the phrase "to convert *unknown unknowns* to *known unknowns*." Our goal is to make our reader aware of important material he or she is not as familiar with as one should be, and to pursue further study to acquire such knowledge.

This is not a programming book. Our goal is to describe the enabling activities that support programmers in producing high-quality software in the context of medical applications. We are less concerned by questions such as 'How should we code?' Our focus, rather, is on answering higher-level questions such as 'How do we decide what we need to code?' and 'How should the process be organized?' There is a wealth of material available that describes the actual coding process, and, therefore, we chose not to duplicate this type of description here.[3]

This is also not a handbook for navigating the regulatory process. Our goal in writing this book was not to create a guide that one can follow to bring a product to market.[4] We aim rather to explain how the regulatory process(es) affects the process of creating software for medical purposes. This will enable both a junior programmer entering this field and, perhaps, a high-level manager supervising a medical software project to understand why things are done the way they are. The regulatory documents and the associated industry standards capture much of what the medical device/software community has learned about this process over the past 20–30 years and constitute primary sources in this field.

Much of the content of this book is an expanded version of handouts prepared for the needs of the class Medical Software Design (BENG 406b) taught at Yale University over the period 2017–2021 by two of the authors (X. Papademetris and A. Quraishi). In our experience, university graduates are badly equipped by their coursework for careers in this general area, as many of the topics that are of critical importance to the medical device/medical software industry are almost never covered in undergraduate curricula.[5]

The Structure of the Book

This book consists of four parts. The first two provide necessary background on the topic. They describe the environment that medical software lives in, be it scientific, regulatory, clinical, managerial, or financial. The third part describes the actual process of design, implementation, and testing of medical software, and the last part presents six case studies in which (for the most part) failure to follow the principles described earlier in the book led to expensive disasters and even deaths.

Part I: Regulatory, Business, and Management Background

This goal of this part of the book is to provide the necessary regulatory and business/management background for medical software. It is subdivided into the following chapters: *Chapter 1: Introduction to Medical Software* provides a brief overview of the entire field of medical software. *Chapter 2: The FDA and Software* focuses on the regulatory aspects of medical software, with a particular emphasis on the regulations and guidance issued by the US Food and Drug Administration (FDA). While this may initially appear to be a purely US-centric approach, it is worth pointing out that the most recent guidance documents from the FDA are based on the work of an international body, the International Medical Devices Regulator Forum (IMDRF), which represents an attempt at establishing international consensus in this field. Furthermore, Section 2.4 within this chapter discusses regulations from the EU and China and other countries.

Chapter 3: Operating within a Healthcare System has two goals. The first is to briefly describe the confusing and complex world of healthcare systems. The second goal is to introduce core clinical information technology standards and to discuss data privacy issues and related legislation, such as HIPAA (US) and GDPR (EU). The chapter concludes with a discussion of the critical topic of cybersecurity.

A quality management system is a core regulatory requirement for the development of safe and high-quality medical software (and any other product). This is described in ISO 9001 [131] and its more specialized derivatives (e.g. ISO 13485 [132]) for software and medical devices. We discuss this topic in *Chapter 4: Quality Management Systems*. Skill and technical competence alone do not suffice; one must perform the work in a properly managed organization whose culture is centered on quality. *Chapter 5: Risk Management* provides a description of this important topic, which is also a core regulatory requirement. Our bases here are the appropriate international standards, in particular ISO 14971 [133].

Medical software development happens in a business environment. *Chapter 6: Taking an Idea to Market: Understanding the Product Journey* is a guided tour of this area with a particular focus on startups and entrepreneurship. We cover here issues related to intellectual property, fundraising, and marketing. Even if the reader is not particular interested in starting his or her own company, these are important issues to understand as they also drive management decisions in established companies.

We conclude this part with a more forward-looking chapter. *Chapter 7: Medical Software Applications and Growth Drivers* provides a review of the current state of the industry and, in particular, describes the excitement and emerging applications based on the use of AI/ML techniques in the area of digital health.

Part II: Scientific and Technical Background

Our aim in this part is to fill in common gaps in background knowledge of undergraduate students (and practitioners) in two important areas.

Chapter 8: Mathematical Background is a fast-paced tour of probability, statistics, and machine learning. Anyone involved in the design of medical software (especially as this touches upon diagnosis and treatment) needs to have some basic background in topics such as detection theory and to understand terms such as false positive and false negative. Most computer scientists, in our experience, have minimal exposure to these topics.

Chapter 9: Topics in Software Engineering introduces basic material from software engineering for those of our readers coming from medical, engineering, and management backgrounds as opposed to computer science. Here, we discuss topics such as software life cycles, software testing, and source code management. All of these are critical in the design and implementation of medical software. We particularly highlight Section 9.3, where we provide a description of the challenges and techniques involved in the use of AI/ML modules within medical software.

Part III: An Example Medical Software Life Cycle Process

This practical application of software life cycles is the heart of this book.[6] Here, we provide a simplified recipe for designing, implementing, testing, and distributing medical software to illustrate the concepts discussed in IEC 62304 [119]. The topic is introduced in *Chapter 10: The Overall Process*.

Next, *Chapter 11: Identifying User Needs* discusses the beginning of a project. This is where we identify what needs to be done and begin to plan how to do it. In this chapter, we also present a sample project, "The Image Guided Neuro-Navigation Project," which we use to anchor our description of the various components of the process in the subsequent chapters.

Chapter 12: The System Requirements Specification discusses the critical process of creating the system requirement document, which is the master plan or anchor for the whole process. Next, *Chapter 13: The Software Design Document* discusses (for the first time) the actual, concrete process of designing software. *Chapter 14: Software Construction, Testing, and Verification* provides a brief description of software implementation and testing.

Chapter 15: Software Validation describes the validation process, which includes a brief description of clinical trials and issues related to the evaluation of AI/ML modules. This part of the book concludes with *Chapter 16: Deployment, Maintenance, and Decommissioning*, a short chapter that discusses these last steps in the software life cycle.

Part IV: Case Studies

This last part of the book consists of six case studies that relate to either software failures or the averting (in the case of the Y2K story) of crises that would have been caused by software issues that were successfully remedied in time. We use these to illustrate and reinforce some of the lessons presented earlier in the book.

Chapter 17: Therac-25 presents the famous case of the Therac-25. These accidents, which were caused by software bugs in a radiation treatment machine, resulted in serious injuries and deaths in a number of cases in the 1980s. They also resulted in the FDA taking an active interest in medical device software. Much of the regulation in this area derives directly or indirectly from the Therac-25 disasters.

Chapter 18: Mars Climate Orbiter describes the loss of the Mars Climate Orbiter. This is a classic failure of integration testing. *Chapter 19: HealthCare.gov: The Failed Launch of a Critical Website* describes the failed launch of the federal health insurance marketplace web page. During the creation of this web page, the planned software process was not followed, resulting in confusion and overruns. *Chapter 20: The 2020 Iowa Caucus App* presents the case of the 2020 Democratic Iowa Caucus App. This is both a case of management failure, effectively trying to do things "on the cheap" in a mission-critical setting, and a case of lack of awareness of the users and their environment.

Chapter 21: The Boeing 737 MAX Disasters describes the ongoing saga involving the design of this airplane. This case, which has some strong parallels to the Therac-25, is an example of an attempt to cut corners and use software to fix/rectify fundamental hardware problems, and using shortcuts and obfuscation to get around regulatory processes. Two 737 MAX airplanes crashed, resulting in hundreds of fatalities.

Our final vignette, *Chapter 22: The Averted Y2K Crisis*, is a success story. This illustrates how proper planning and thorough work averted what might have become a serious problem caused by legacy computer software design.

Target Audience

Undergraduate students in computer science and engineering: The initial stimulus for writing this book was to address the needs of our undergraduate students in the Yale class Medical Software Design (BENG 406b).

Junior software engineers in the medical device industry: Our hope is that Part I of this book will provide this group with an understanding of the environment they find themselves in and some justification as to why things are done the way they are. Part II may also be profitable to those who lack some of the technical background. Their company will have a specific process for the actual design phase, so Part III may be a little less applicable, but may also, again, provide some explanations as to why their company's process is what it is. One of the reviewers of this book also suggested that knowledge of this material will be helpful in the interview process when applying for jobs in the medical device industry.

Senior management and/or policymakers supervising and/or managing software projects: Such individuals need to be able to assess the quality of the underlying software and to account for the risks that aspects of this software pose to their own decision-making process. For these individuals, the primary utility of a book such as this one (especially Parts I and II) may be to enable them to ask the right questions of their subordinates, who are actually directly managing the software process. The case studies (Part IV) may also provide sobering reading.

Those interested in starting a digital health company: This new entrepreneur in digital health is entering the minefield of the medical world, which comes with regulatory constraints (e.g. the FDA) and environmental constraints (hospitals) that make the process significantly harder (at least at the beginning). Our hope is that this person will be able to quickly skim this book and get a sense of what their task will involve and how to seek appropriate help (e.g. consultants) as they begin this challenging (and potentially highly rewarding) task of setting up a company (e.g. quality management systems – see Chapter 4). Chapter 6 may provide them with a useful road-map for the business aspects of this task, including coverage of topics such as intellectual property, fundraising, and setting up human studies/clinical trials. Chapter 7 reviews the current state of the industry and provides examples of recent work that may serve as an inspiration for new entrepreneurs. Finally, reading through the software process part of the book (Part III) will at the very least make it plain that they must document their software designs before they start coding if they would like anything they produce to meet regulatory standards.

NOTES

1. These regulatory documents [64, 69, 70, 82, 89, 91, 92, 125–127, 188] and industry standards [1, 2, 119, 122–124, 131–134] form a good initial list.

2. Probably at the top of this list should be the book *Medical Device Software Verification, Validation, and Compliance* [261] by D.A. Vogel. Though slightly dated, it contains much valuable information.

3. In teaching a class on medical software we have consistently emphasized that "figuring out what to code is much, much, harder than actually coding." We, obviously, recognize that, in certain cases, the actual coding process can be highly complex and requires significant expertise and experience! There are many other classes that provide training on the coding process itself.

4. The regulatory process differs from country to country and it is critical to ascertain what the requirements are for any particular country in which approval is sought. The good news is that there are ongoing efforts to harmonize these regulations through the work of the IMDRF.

5. In Chapters 11–13 there is a specific section that applies the material in that chapter to our fictional image-guided navigation software project presented in Chapter 10. The hope is that this provides students with concrete examples of how to go about completing class assignments. We also provide sample class assignments in these chapters, and also in Chapter 15. One of the authors was involved in the design and implementation of a research interface for the BrainLAB system [198], but he had no (and still does not have a) financial connection to the company. This was a purely scientific collaboration as part of an NIH-funded project.

6. When taking a class on medical software, many students expect the class to start at this point. By contrast, many of the regulations highlight the importance of quality management systems and risk management in addition to the software life cycle process.

Acknowledgments

We would first like to thank the students, the teaching fellows, and the mentors who participated in BENG 406b over the past few years. They were the "beta-testers" of the book and we thank them for their patience through the years and their helpful suggestions for improving the manuscript.

We would like to thank our anonymous reviewers for their kind and helpful feedback. In addition, we would like to thank the friends and colleagues who volunteered to read portions of this book and offer helpful comments. In alphabetical order: Tina Kapur, Sean Khozin, Phan Luu, John Onofrey, Nicholas Petrick, Steve Pieper, Saradwata Sarkar, Lawrence Staib, Don Tucker, and Rajesh Venkataraman. An Qu, postgraduate associate at Yale, double-checked the unofficial English translation of one of the Chinese regulatory documents [189]. Robert Bell took part in the many early discussions about this book with Drs. Papademetris and Quraishi. We would also like to thank the anonymous peer reviewers of the book for their many constructive comments.

We would like to thank our colleagues at Yale biomedical engineering, in particular Mark Saltzman, who had the idea for creating this class and provided the institutional support and mentorship as it was being set up.

Ellie Gabriel, a Yale biomedical engineering undergraduate, wrote the initial drafts of the six case studies in Part IV of this book as part of a summer internship (2020) she did under the supervision of Dr. Papademetris. She also created the illustration for Figure 10.4. She, together with Maria Papademetris, did much of the initial proofreading for the text.

We would also like to thank our editors at Cambridge University Press, Elizabeth Horne and Jane Adams, and our Content Manager, Rachel Norridge, for their help in shepherding us through the process in the middle of a global pandemic. Without their help, this would have been a much inferior manuscript.

Part I

Regulatory, Business, and Management Background

1   Introduction to Medical Software

INTRODUCTION

This chapter begins by defining what medical software is and what makes it unique, and describing the regulatory process that governs it (Section 1.1), including a brief introduction to industry standards. Following this, we discuss the constraints (both business and technical) placed on the software process by the medical environment (Section 1.2). The chapter concludes with a discussion of the challenges that result from the use of artificial intelligence/machine learning (AI/ML) techniques (Section 1.3). This section also provides references to other parts of this book where we discuss AI/ML issues.

1.1   Medical Software and the Regulatory Process

1.1.1   What is Medical Software?

Medical software is software that is used in a clinical setting, either as part of a medical device or as a standalone application. An example of the former is software that manages the delivery of radiation in a radiation treatment machine in a therapeutic radiology setting (e.g. lung radiotherapy). In this case, the software controls a device, whose misuse can cause serious harm to the patient. This example is not chosen at random! The reader may recognize that this is precisely the setting of the Therac-25 [154–156] incidents in the 1980s that precipitated the involvement of the Food and Drug Administration (FDA) in regulating medical software. We discuss the case of Therac-25 in more detail in Section 2.1.5 and the Vignette presented in Chapter 17.

At the other extreme, we may have software for analysis of patient medical images by a radiologist for diagnostic purposes. In this second scenario, the danger to the patient is potentially less serious, as the software does not interact with the patient directly. However, an error in the analysis can cause an error in the diagnosis, which may also lead to harm caused by using the wrong treatments later in the process. This is a pure software device (no hardware), formally known as SaMD – software as a medical device.

Interestingly, a recent guidance document from the Health Sciences Authority of Singapore [234] divides software that is not part of a device into three categories: (1) standalone software, (2) standalone mobile applications, and (3) web-based software. This reflects the additional challenges posed by mobile and web-based applications.

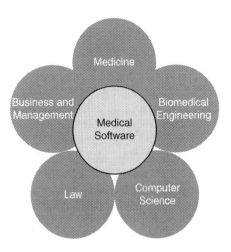

Figure 1.1 Medical software and related fields. Medical software lies at the intersection of several different fields and requires teams with expertise in these very different areas. This textbook attempts to provide, in a single relatively brief volume, an introduction to this broad area.

In our view, medical software is a field that lives in the intersection of five areas: (1) medicine, (2) biomedical engineering, (3) computer science (software engineering), (4) law, and (5) business and management. This is illustrated in Figure 1.1.

1.1.2 Some Unique Aspects of Medical Software

In the United States, most medical software (some software is exempt, as discussed in Section 2.1.1) and medical devices require clearance or approval by the FDA prior to use in patient care.[1] Hence, the usual software development process of (1) design, (2) implement, (3) test, and (4) market has an additional step prior to "market," which is to *obtain regulatory clearance.* As an aside: the terms "clearance" and "approval" are used for different types of regulatory processes – see Section 2.3 for more details. We will sometimes use the terms "approved"/"approval" loosely to indicate either of these processes, that is, that the product is "blessed" by the FDA to be used for patient care.

This makes medical software design and development a regulated activity that must be performed in accordance with specific laws passed by Congress and regulations issued by the FDA.[2] Therefore, while in some software work the explicit writing of system specifications is considered a good idea but not mandatory,[3] in the medical realm written specifications are explicitly required. Specifically, the FDA's Quality Systems Regulations (QSR) [69] (Section C – Design Controls) states in no uncertain terms that: "The design input requirements shall be documented and shall be reviewed and approved by a designated individual(s)."

In general, designing medical software in a way that meets regulatory criteria requires more than validation of the completed software. The structure and the environment in which the software development takes place are also critical, as discussed in Chapter 4 on quality management systems. We must be able to establish procedures and plans that demonstrate that the overall software development process is "controlled" and follows an actual plan rather than ad-hoc programming.

As a helpful analogy, consider the case of an "FDA-regulated" restaurant. The FDA is also interested in the process used to develop the food (software), not just whether it is safe (and presumably tastes good). This means that the restaurant needs to be organized in such a way that good and healthy food is likely to be made. This includes having formal processes for selecting recipes, training staff, testing for quality and how the staff goes about cooking and serving the food, and managing risk. The FDA will want to know whether the food is safe and tasty and be sure that the utensils will not cause harm to the customer (patient). Finally, the FDA can inspect a restaurant at any time to verify compliance (not just taste the food!). International standards (see Section 1.1.4) such as ISO 9001 [131] are increasingly placing more emphasis on the organization as a whole rather than narrowly focusing on the individual product.

1.1.3   The FDA and Software

The FDA's involvement with software began with the Therac-25 incidents in the 1980s, when a software bug resulted in the delivery of excess radiation to patients by a radiotherapy treatment device, leading to severe injuries and deaths [154, 156].

The classic FDA guidance document on software is titled *General Principles on Software Validation* (GPSV) [70]. The GPSV describes at length the FDA's experience with software and its suggestions (which, while not binding, one would be wise to treat as more than mere recommendations) for how software should be developed and validated. We will discuss this in more detail in Section 2.2, but one point is particularly worth highlighting in this introductory chapter. After a very interesting discussion of the differences between software and hardware, the GPSV states: "For these and other reasons, software engineering needs an even greater level of managerial scrutiny and control than does hardware engineering." This should be eye-opening to those (especially at the managerial level) who think of software as easier or less critical than hardware. The FDA's experience of the matter suggests the opposite.

More recently, the FDA and the International Medical Devices Regulators Forum (IMDRF) have created additional guidelines specifically for the case of standalone software under the umbrella term of SaMD [92, 125–127]. Such software is becoming increasingly common (see Chapter 7).

1.1.4   Industry Standards

While the FDA and its counterparts in other countries are the legally responsible regulatory bodies that clear/approve software for medical use, much of the specific requirements for structuring the organization, project management, software design, implementation, verification, and validation derive from industry standards created by organizations such as the International Standards Organization (ISO) [135]. As stated on its website, the ISO is an independent nongovernmental international organization based in Geneva, Switzerland. Its membership consists of 164 national standards bodies, including ANSI – the American National Standards Institute. The ISO was founded in 1946 [135] with the participation of 25 countries. It produces standards on a variety of topics such as (1) quality management systems, (2) environmental management systems, (3) health and safety standards, (4) energy management standards, (5) food safety standards, and (6) IT security standards. A standard is defined on the ISO website as "a formula that describes the best way of doing something … Standards are the distilled wisdom of people with expertise in their subject matter and who know the needs of the organizations they represent – people such as manufacturers, sellers, buyers, customers, trade associations, users or regulators."

These standards represent sector-specific best practices. They are written by industry workgroups that include representatives from manufacturers and often regulators.

Relation between Regulations and Standards

While compliance with such standards is voluntary, many regulators recognize compliance with specific standards as evidence of "proper work." For example, consider the case of the standard IEC 62304, titled "Medical device software – software life cycle processes" [2, 119]. In a document from the IMDRF [128], "Statement regarding use of IEC 62304:2006 'Medical device software – software life cycle processes'" [128], one can find statements from different member organizations (i.e. regulators from different countries) relating to the applicability of these standards. To quote two examples, the first from the US FDA and the second from the China Food and Drug Administration (CFDA):

IEC 62304:2006 is recognized by the US FDA medical device program as a consensus standard for which a person may submit a declaration of conformity in order to meet a premarket submission requirement or other requirements to which a standard is applicable. US FDA by recognizing IEC 62304:2006 is acknowledging that the process activities and tasks identified in this standard when used with a good quality management system and risk management system can help assure safe design and maintenance of software used in medical devices.

The IEC 62304:2006 had been translated into China industry standard: YY/T 0664-2008 equally and implement from 2009.6.1, it isn't mandatory standard, and just is recommended standard.[4]

To conclude, we will quote IEC 62304 [119] on the details of medical software. The document's introduction section states:

As a basic foundation it is assumed that MEDICAL DEVICE SOFTWARE is developed and maintained within a quality management system (see 4.1) and a RISK MANAGEMENT system (see 4.2). The RISK MANAGEMENT PROCESS is already very well addressed by the International Standard ISO 14971.

This completes the loop back to the US FDA statement above. IEC 62304 effectively says that you need to follow these best practices in your software process, but it assumes the presence of a quality management system (we discuss this in Chapter 4) and a risk management system (see Chapter 5). IEC 62304 will be our guide and anchor in Part III of this book, where we will discuss the actual process of medical software design, implementation, and testing.

In general, while adherence to these standards is voluntary, many regulators recognize specific standards as representing appropriate/best practices, so a company that adheres to specific standards is likely to have an easier path to obtaining regulatory clearance. In addition, many specific regulations are explicitly based on industry standards. As an example, the IMDRF document on quality management systems [127] makes explicit and copious references to the industry standard ISO 13485 [132]. This is not surprising, as (1) regulatory organizations often participate in the workgroups that write the standards, and (2) the *real* domain expertise in particular topics often lies in those organizations that have significant experience in producing such devices.

Important Standards for Medical Software

For medical software, the key standards are probably:

- *ISO 9001: Quality management systems* [131];
- *ISO 13485: Medical devices – quality management systems* [132];
- *ISO 14971: Application of risk management to medical devices* [133];
- *IEC 62304: Medical device software – system life cycle processes* [119]; and
- *IEC 62366: Medical devices – application of usability engineering to medical devices (Parts 1 & 2)* [122, 124].

While most of these standards come through the ISO, other organizations also issue standards. These include the Institute of Electrical and Electronic Engineers (IEEE), the International Electrotechnical Commission (IEC), and the Association for the Advancement of Medical Instrumentation (AAMI).

For those interested in Agile methodologies the AAMI technical information report (TIR) 45, titled "Guidance on the use of AGILE practices in the development of medical device software" [1], should also be consulted (see also Section 9.2.2).

In this book we will refer explicitly to (and quote) the regulatory documents more than the industry standards. This is primarily because the former are freely available and, as such, are more easily accessible to the average student. Naturally, those readers who have significant responsibility in a medical device/medical software company should read the actual standards themselves and probably attend more advanced training sessions on how to apply them.

Data Privacy and Cybersecurity

In addition to the "standard" regulatory documents, one needs to understand the implications of the increasingly strong rules on data privacy and security as a result of regulations such as HIPAA [49] in the United States and more recently GDPR [65] in the EU. Data privacy is critical for most medical software and needs to be accounted for at the beginning of the process rather than at the end. A related issue is cybersecurity. We discuss these topics in more detail in Section 3.4.

1.2   The Medical Domain and Software

1.2.1   The Business Environment for Medical Products

The development of a new product can be divided into research and development (R&D). From the perspective of regulated medical software development, research can be defined as unregulated exploration of the problem. Development, on the other hand, is the regulated process of product design and construction of the solution [69, 261]. In the context of medical software development, the research phase involves the development and testing of new technology such as ML algorithms and visualization techniques. Thus, the research through which we discover the requirements for the final product is outside the regulatory process; however, much like basic scientific research, it should still be documented carefully. The development phase is the process of taking this new technology, combining it with other standard tools, and creating a new software package for patient care. It starts once the requirements for the product have been finalized.

It is important for the software designer/engineer to have a sense of the underlying business environment in which their product will live or die. Given the complexity and cost of medical products, it is worth, at the start of the project, mapping out the pathway for the project from idea to product and identifying the risks involved ("failure modes" in engineering terminology) in this path and how the product may fail.

Figure 1.2, which comes from the MIT hack medicine philosophy of healthcare entrepreneurship, illustrates these risks. Here, we plot value (vertical axis) vs. development time (horizontal axis). The key lesson from this figure is that while most companies focus on starting with the technical risk, it is most important to start with de-risking the "market risk" by really testing and validating the first customer(s) that will pay for the service or product. Quite often there is an assumption that "someone" wants the technology being created, which turns out later to be untrue – this model attempts to minimize that risk first and foremost and to continue this during the technology development process. This can be done by properly assessing user needs and establishing the right specifications for the project. Once the needs have been identified, the sequence suggests that the first mode of failure is "technology risk," which simply means that our proposed new idea (research phase) does not work out as intended. The next step is "management risk," which refers to a failure to manage the development of the new technology into a product with a defined customer; this may involve organizational failure or financial failures – that is, failure to raise enough money to support the process. Next, we have regulatory risk – our new product fails to be cleared by the regulatory agencies despite appearing to work satisfactorily. The cause of this might be a failure in the process and is best handled by designing the product from the ground up in a manner compliant with regulatory guidelines. Finally, we can have a fully working, regulator-cleared product that fails in the market because nobody is interested in paying for it. At the risk of repeating ourselves: The regulatory agencies are interested in the whole process, not just the final output, so the process of obtaining regulatory clearance starts with the initial design of the software as opposed to just its validation.

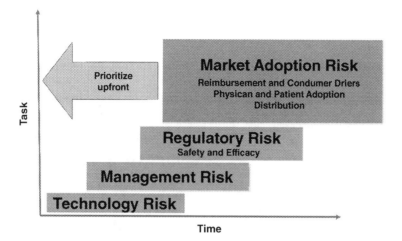

Figure 1.2 Product development and risk management. This figure shows the different aspects of risk management for a project, ranging from technology risk (will it work?) to management risk (can we pull this off?) to regulatory risk (will the FDA allow this?) to, finally, market adoption risk (will anybody buy this?). This figure comes from a presentation by Zen Chu. Z. Chu. Hacking medicine toolkit. Grand Hack presentation, April 2019. Used with permission from the author.

Failure from market risk, therefore, is very expensive after the technology has been developed and an effort made to obtain regulatory clearance. Despite this, many medical software products are created without the designers speaking to potential customers. We will discuss these topics at length in the rest of the book.

1.2.2 The Clinical Information Technology Environment

Some software is completely standalone and does not need to communicate with the outside world in any way.[5] For example, a calculator application is completely self-contained and does not need to import or export data from other software packages/servers (other than perhaps for supporting the operating system's global clipboard copy-and-paste operations). While this is sometimes the case for medical systems, it is far more common that any given piece of medical software plays a role in a larger process in which the inputs may come from an electronic patient record database [14] or the clinical imaging database [208] (see section 3.3 for a description of these databases.) The software may then perform some manipulation, creating some form of output, which may also be stored in an external database for later use. The interoperability standards that govern the communication between such pieces of software create "environmental constraints" that need to be properly understood as part of the software design process. Ignoring these may result in a great piece of software that is incapable of functioning as part of a complex workflow and is, hence, ultimately useless for the end user.

1.2.3   Understanding a Clinical Problem and User Needs

Successful design in healthcare starts with the clinical pain point that is felt either by the patient/caregiver, the provider, or the healthcare delivery mechanism; only through clinical pull can projects be effectively created to properly meet user needs. The software designer must understand the bigger context in which the need arises. In general, one must first understand (1) the underlying disease/clinical situation, (2) the current methods used to address/treat the problem if any exist, and (3) the limitations of these methods and how they cause problems to users. This will allow one to establish the potential impact of any proposed solution and to assess whether the project is worth carrying out.[6]

Understanding the needs of the user will involve spending a significant amount of time observing clinical procedures and asking questions to make sure the designer truly understands the context and the needs of the user.[7] One needs to learn the "vocabulary" of the users and to speak some of "their language" to be able to communicate with them in a meaningful way. A key danger is that most physicians or clinical providers are busy, so time may be limited. However, failure to properly understand the users' needs will result in a flawed product, probably produced at great expense. Sometimes many of the users' needs are so obvious to them that they may neglect to mention these needs. The job of the software designer (and/or the business analyst in a larger company) is to make the users' implicit needs explicit. We discuss this topic in more detail in Chapter 11.

Finally, one needs to think hard about how to validate the proposed solution. What are the performance metrics (ideally quantitative) that can be used to demonstrate to the users' satisfaction that the software actually does what it is supposed to do? In our experience, this is a concept alien to most physicians (who can tend to think, "I will know if it works when I use it"), but such metrics are useful for proving efficacy in the regulatory process (and for marketing, etc.).

1.2.4   Software Engineering and Medical Software

What does the process of successfully designing medical software that will meet regulatory standards actually look like?

The first and critical step, worth reiterating here, is that the process must happen within an organization that has an appropriate quality management system (QMS) [127, 131, 132, 134] for this purpose. This system forms the critical environment/substrate in which the software will be designed, developed, and produced. In the absence of this environment, one cannot (at least according to the regulatory and industry consensus – an example is the comment to this effect in IEC 62304 [119],

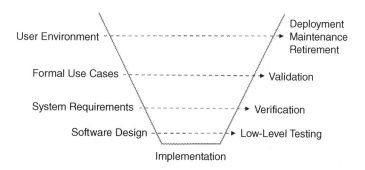

User Environment

Formal Use Cases

System Requirements

Software Design

Implementation

Deployment
Maintenance
Retirement

Validation

Verification

Low-Level Testing

Figure 1.3 The V Model. We present here an adaptation of the V Model for the purposes of this introductory book. This schematic will serve as a roadmap for Part III of this book. The V Model was originally proposed by Forsberg and Mooz [95].

which is quoted in Section 1.1.4) create quality software that demonstrates efficacy, safety, and security. First-time employees in the medical device industry are often surprised to discover that, before they are asked to write a single line of code they are trained (one might say "indoctrinated") in their company's QMS processes.

In this textbook we organize our discussion of the medical software life cycle process (Part III) around the so-called *V Model* [95], which is in general agreement with the structure presented in the international standard IEC 62304 [119]. The V Model is widely used in systems engineering. We show an adapted version of this model in Figure 1.3, following the terminology/structure proposed in the IMDRF QMS document [127]. The V Model pairs "creation" and "use/testing" steps. The process begins with observing the users' environment. The corresponding elements on the other side are those processes in which the finished project interacts with the user (deployment, maintenance, retirement). The next step is formal establishment of user needs requirements, and the corresponding element is *validation*, the process by means of which we ensure that our software satisfies the needs of our users. The next step is system requirements, in which the system is designed (at a high level) to meet the requirements (both user-derived and regulatory/standards-derived), and the corresponding step is *verification*, the process of ensuring that the software was correctly built/designed. The last descending step is software design, the design of module/unit-level functionality to satisfy the system requirements; the corresponding step on the ascending line is software testing, the creation of low-level unit/module tests to ensure that the software functionality at this level was implemented correctly. At the bottom of the "V" is the actual software implementation work.

Risk (i.e. the possibility of causing harm to a patient or a caregiver – see Chapter 5) management takes place throughout this process, as issues can be introduced at all stages of the life cycle. This includes a careful evaluation of software usability (see Section 13.5) and cybersecurity (see Section 3.5).

1.3   Emerging Challenges from the use of Artificial Intelligence/Machine Learning Techniques

At the time of this writing, there is a huge explosion of interest (and hype) in the use of AI/ML techniques in medicine. Such techniques are at the heart of the emerging field of digital health (see Chapter 7) and promise to revolutionize the design and development of medical software. They are especially powerful when combined with mobile device technologies. While formal definitions of these terms vary, the FDA draft guidance document [82] cites the following definition of AI:

[T]he science and engineering of making intelligent machines, especially intelligent computer programs [168]. AI can use different techniques, such as ML, to produce intelligent behavior, including models based on statistical analysis of data, and expert systems that primarily rely on if-then statements. In this paper, we refer to an ML system as a system that has the capacity to learn based on training on a specific task by tracking performance measure(s). AI, and specifically ML, are techniques used to design and train software algorithms to learn from and act on data.

The biggest technological driver in this area has been the development of deep learning techniques. Deep learning is a form of machine learning that uses deep neural networks (networks that have a large number of layers) as the learning mechanism. A good overview of the field can be found in the book *Deep Learning* [100] by Goodfellow et al. A recent review paper by Esteva et al. [61] surveys the use of deep learning in healthcare.

We will discuss topics related to AI/ML throughout this book. The most important sections are as follows:

- We present an overview of the current regulatory environment in Section 2.6.
- We describe the core mathematical principles of ML in Section 8.4.
- We present software engineering techniques for AI/ML methods in Section 9.3. There is a major paradigm shift here that involves a primary focus on data management.
- A brief discussion of risk assessment for AI/ML modules can be found in Section 5.2.4.
- We briefly discuss practical planning issues relating to the use of AI/ML in Section 10.5.3.
- We cover issues relating to the validation of ML/AI-aided software in Section 15.1.2 – this is a continuation of Section 9.3.

While international standards in this area are still under development,[8] there are a number of documents from national regulators and standards organizations that will guide our discussion. The main ones we consulted in writing this book are:

1. US FDA: Proposed regulatory framework for Modifications to Artificial intelligence/machine Learning (AI/ML)-based software as a medical device (SaMD) – discussion paper and request for feedback [82];
2. China NMPA: "Technical guideline on AI-aided software" [189];

3. A summary of current thinking from the Science Board of the Pharmaceuticals and Medical Devices Agency, Japan (PMDA), in a recent review paper by Kiyoyuki et al. [31];

4. The German Institute for Standardization's "*DIN SPEC 92001-1:2019-4: Artificial intelligence – life cycle processes and quality requirements – part 1: quality meta model*" [53];

5. A guidance document titled "Review and approval of artificial intelligence (AI) and big data-based medical devices" [215] from the South Korean Ministry of Drug and Safety; and

6. A document on AI testing jointly produced by the Korean and Chinese Software Testing Qualifications Board [149].

There are many other guidance documents and position papers in this area. Most of these are written at a very high level and are not particularly useful for practical purposes. The Johner Institute [140] has a useful list of such documents on their webpage that significantly aided our research in writing this book. Of these, one that we have found useful is the white paper produced by Xavier Health [228], which focuses on issues related to continuously learning systems.

1.4   Conclusions

Writing medical software involves a lot more than the writing of the software itself. It requires an understanding of the constraints imposed by the regulatory process, the medical environment, and the overall business environment. These constraints directly impact how the software is created. A lack of appreciation for these constraints can cause an otherwise promising idea to fail. We will revisit these topics extensively throughout the rest of the book.

1.5   Summary

- Creating medical software is a regulated activity. Most medical software requires regulatory review before it can be used for patient care, and this imposes constraints as to how the software is designed, implemented, and validated.
- Agencies such as the FDA (United States) and the European Medicines Agency (EU) issue regulations and guidance documents to guide this process. Increasingly, these documents are being harmonized under the auspices of the IMDRF.
- Another important source of guidance is the international standards produced by organizations such as the ISO. These represent industry best practices and are commonly recognized by regulatory agencies.
- The constraints created by both the business environment and the clinical information technology environment must be accounted for in designing and developing software. It is critical that the designer understands the needs of the user(s) and the

environment in which they need to work (including interoperability with existing systems).

- Developing medical products is an expensive process and business considerations must be taken into account.
- There are new challenges and opportunities arising from the increasing use of AI/ML techniques that are impacting both software designers and regulators.

RECOMMENDED READING AND RESOURCES

For those working in the medical software/medical device industry, the industry standards mentioned in this chapter are indispensable. (These unfortunately are expensive and, as such, are not easily accessible by students.) One should probably start with these three:

ISO:13485 Medical devices – quality management systems – requirements for regulatory purposes, 2016.

ISO:14971:2019 Medical devices – application of risk management to medical devices, 2019.

IEC 62304 Medical device software – software life cycle processes, May 2006.

NOTES

1. The situation in the EU is similar; CE Mark certification plays the same role as FDA clearance in the United States.
2. This is the case for the United States. The situation in other zones is similar. In the EU, there are laws passed by the European Parliament and regulations issued by the European Medicines Agency.
3. For example, consider the statement: "The requirements do not have to be written, but they have to become known to the builders" from the well-known book on software requirements by Robertson and Robertson [216] – see Truth #5 in chapter 1 of that book.
4. The CFDA was later renamed the National Medical Products Administration (NMPA).
5. This scenario was far more common in the days before the development of the Internet and associated cloud services.
6. Another key concern is to determine who is getting paid to provide the current care and whether our plan is to cooperate (improve the current system) or compete with them.
7. A potential pitfall here is that often the physicians or clinical providers driving the development of new products are highly skilled and experienced practitioners working at high-end academic hospitals. Their needs and environment may be different from those of their less-experienced colleagues working in community hospitals.
8. One example is *ISO/IEC CD 23053.2, Framework for artificial intelligence (AI) systems using machine learning (ML)*.

2   The FDA and Software

INTRODUCTION

This chapter describes the regulatory process for medical software, with a particular emphasis on the documents issued by the United States Food and Drug Administration (FDA). We first describe the FDA itself (Section 2.1), including a brief history of how the current process has evolved over the past century. We then review in detail some key FDA regulatory documents (Section 2.2) and provide an overview of the actual regulatory process (Section 2.3). After this, we survey related documents and agencies from the rest of the world (Section 2.4). Unsurprisingly, given that most regulatory agencies are participants in the International Medical Device Regulators Forum (IMDRF), there is significant convergence in the regulations operating in different countries. Section 2.5 discusses the implications of these regulations for the software process. The chapter concludes with a look at emerging regulatory guidance on the use of artificial intelligence/machine learning (AI/ML) techniques in medical software and some comments about the applicability of some of the concepts in the medical software regulations to other industries (Section 2.6).

2.1   The FDA and Medical Devices

2.1.1   Introduction and Mission Statement

The FDA is an agency of the United States federal government (part of the Department of Health and Human Services). It is responsible for protecting and promoting public health through the control and regulation of medical devices, as well as many other products such as prescription and over-the-counter (OTC) drugs, vaccines, and blood transfusions, as well as cosmetics, food, tobacco, veterinary, radiation-emitting devices, and other products. The mission of the FDA is, in part:

The Food and Drug Administration is responsible for protecting the public health by ensuring the **safety**, **efficacy**, and **security** of human and veterinary drugs, biological products, and medical devices; and by ensuring the safety of our nation's food supply, cosmetics, and products that emit radiation. (Emphasis added.)

Following Vogel [261], we define the three key words (in bold) as:

- **Efficacy:** the software must be effective in achieving the claims made for its intended use.[1]
- **Safety:** use of the software does not expose the patient or the user to any undue risk from its use.
- **Security:** the software was designed to protect patients and users from unintentional and malicious misuse.

Simply put, your software must work as advertised, must not cause harm (usability evaluation is critical here – see Section 13.5), and cannot be abused to create harm to either users or patients (e.g. cybersecurity issues – see Section 3.5). The FDA uses a benefit/risk analysis for approvals to ensure a product's benefits outweigh the risk of use.

The FDA only regulates software that is classified as a medical device. It is important to note, therefore, that not all medical software is subject to FDA regulation as not all software used in the medical world is a medical device. As the FDA itself states [82]:[2]

as detailed in section 502(o) of the FD&C Act, software functions intended (1) for administrative support of a health care facility, (2) for maintaining or encouraging a healthy lifestyle, (3) to serve as electronic patient records, (4) for transferring, storing, converting formats, or displaying data, or (5) to provide certain, limited clinical decision support are not medical devices and are not subject to FDA regulation.

While some applications are not currently regulated (e.g. wellness [78] and clinical decision support [76]), it is still good practice to follow the guidelines (e.g. quality management systems, or QMS) set out by the regulatory agencies for the development and evaluation in the development of such software. There are two reasons for this: (1) what is regulated may shift over time, and what were previously unregulated apps may end up being regulated; and (2) following these best practices is also an excellent form of liability protection in the case that something goes wrong and a user experiences harm from the use of such an unregulated app.

2.1.2   Structure and Operation of the FDA

The FDA has six main centers, each of which is tasked with oversight of different medical and non-medical areas. Medical software is under the jurisdiction of the FDA Center for Devices and Radiological Health (CDRH), which is responsible for premarket approval of all medical devices, as well as oversight of manufacturing and performance. The CDRH also oversees monoclonal antibodies and radiation safety for all radiation-emitting devices, such as those used in medical imaging, as well as home and commercial products such as cellular phones and microwave ovens. Three other FDA centers that are responsible for medical products are: the Center for Drug Evaluation and Research (CDER), which approves and monitors most drugs and pharmaceuticals; the Center for Biologics Evaluation and Research (CBER), which reviews and tracks biologics such as therapeutic proteins, vaccines, gene therapy, and cell, tissue, and blood products; and a number of offices such as the Oncology Center of Excellence (OCE), which focuses on advancing the development and regulation of "products for patients with cancer."

The different FDA centers issue their own guidelines, and each sets requirements for product approval as well as review of safety and effectiveness. Submissions to the CDRH can be evaluated through different pathways depending on novelty, complexity, and safety. Products are first classified based on risk and regulatory controls (see Section 2.3). Premarket Approval Applications (PMAs [81]) have stringent requirements because these devices are expected to "support or sustain life," or the particular type of product has not been previously evaluated. The PMA should include data from pilot, feasibility, and pivotal human clinical trials that compare the device to a control group, often using the current standard of care. These studies and their impact on software development will be described in further detail in Section 2.3. Alternatively, the 510(k) [86] pathway is considered somewhat less rigorous because when using this process the investigational product is shown to be as safe, as and substantially equivalent to, another legally marketed device or product. There is also the De Novo classification process which allows for novel devices (and software) to be approved without the full PMA process if one can "provide reasonable assurance of safety and effectiveness for the intended use, but for which there is no legally marketed predicate device" [77].

It should be noted that drugs and biologics are reviewed by CDER and CBER through different submission processes called New Drug Applications (NDA) or Biologics License Applications (BLA). These products typically have gone through multiple stages of laboratory testing, pre-clinical animal studies and human clinical trials,[3] although there are some exceptions. Approval requirements for drugs and biologics include human investigations, which are divided into Phase 1 safety, Phase 2 dosing, and Phase 3 efficacy studies, all of which can take up to 10 years and may cost hundreds of millions of dollars. Post-approval safety monitoring and "real-world" performance is conducted as "postmarket" studies. NDA and BLA submissions can be substantial and must also contain detailed pharmacokinetics of the product, effects on physiology, accumulation, and elimination, as well as manufacturing specifications and controls.

2.1.3 Origins of the FDA

The FDA began in 1906[4] with the passage of the Federal Food and Drugs Act, much of which came in response to conditions in the meat-packing industry and the addition of poisonous food additives. Approximately 30 years later, Congress passed the 1938 Food, Drug and Cosmetic Act in the wake of the Elixir Sulfanilamide disaster (1937). This was a mild antibiotic that was specifically marketed to pediatric patients. The solvent in this product, however, was a highly toxic chemical analog of antifreeze.

The use of this drug resulted in the deaths of over 100 people. The 1938 Act expanded the FDA's authority from food inspection to cosmetics and therapeutic devices and prohibited false therapeutic claims in advertising. The Act also allowed the FDA to perform factory inspections.

The history is naturally marked by a number of public health challenges. The next landmark was the thalidomide disaster in 1962. Thalidomide was marketed by the German firm Chemie Grunenthal as a mild sleeping pill safe even for pregnant women. It was sold in Europe, but the FDA managed to keep it out of the US market [170]. It was associated with the births of thousands of malformed babies in Western Europe. As a result of this, Congress passed in 1962 the Kefauver–Harris Drug Amendments to the Federal Food Drug & Cosmetics Act. This added the requirement that prior to marketing a drug, firms had to prove not only safety but also evidence of effectiveness for the product's intended use.

2.1.4   The FDA and Medical Devices

Another major landmark came with the Dalkon Shield disaster. This was an intrauterine contraceptive device which, by 1974, was used by approximately 2.5 million women and whose use led to serious inflammation and death [111]. As reported in the *New York Times* [147]:

The Dalkon Shield – an intrauterine contraceptive device – is gone from the market now, but not forgotten. As many as 200,000 American women have testified that they were injured by the device and have filed claims against the A.H. Robins Company, which sold it to an estimated 2.5 million women during a four-year period in the early 1970's. Robins has filed for bankruptcy and is currently engaged in legal arguments over whether the $1.75 billion it proposes as a compensatory fund to women who claim to have been injured by the shield is anywhere near enough.

As a result, the law was modified in 1976 (the Medical Device Amendments) to require testing and approval of medical devices for the first time. The Act (see the FDA website) established three regulatory classes for medical devices, ranging from Class I (low risk) to Class III (high risk) [74].

2.1.5   The FDA and Software: The Therac-25 Tragedy

The Therac-25 was a radiation treatment machine made by Atomic Energy of Canada Limited (AECL) in 1982. This was a successor to the earlier Therac-6 and Therac-20 units, which had been made in collaboration with a French company, CGR. Such devices aim to treat cancer by targeting radiation beams at tumors. The treatment plans take advantage of the fact that normal cells recover faster than cancer cells from radiation treatment. So rather than delivering a massive dose at once that will kill both normal and cancer cells, most treatment plans apply multiple (in the case of prostate cancer as many as daily doses over 5–6 weeks) smaller doses over time, allowing for the normal tissue to recover, thus reducing complications.

The Therac-25 was designed to deliver multiple types of radiation for different types of tumors. Due to a bug in the software, the machine delivered massive doses of radiation to patients (at least six cases are known), resulting in serious injuries and deaths. This was one of the first cases in which software was directly implicated in causing serious harm to a patient, which resulted in the FDA beginning to regulate software.

A comprehensive description of this set of incidents can be found in Appendix A of Nancy Leveson's book *Safeware* [154].[5] Our focus here is simply on the lessons for software design and how this affected the FDA as this disaster led, for the first time, to the explicit regulation of medical device software.

2.1.6 Lessons from Therac-25

Leveson [154] summarizes the causal factors in the accidents as follows:

- overconfidence in the software;
- confusing reliability with safety;
- lack of defensive design;
- failure to eliminate root causes;
- complacency;
- unrealistic risk assessments;
- inadequate investigation or follow-up on accident reports;
- inadequate software engineering practices;
- software reuse; and
- safe vs. friendly user interfaces.

We will not discuss all of these issues at length here, but we will pick up the thread of many of these aspects in the following chapters. Almost all of the issues seem obvious,[6] with one key exception: software reuse. Software reuse is a key "good" component of most solid software development. Once we have a solid, tested version of some functionality, reimplementing it from scratch for a new project is generally (and correctly) frowned upon; hence, it is a little surprising to see software reuse as a causal factor. The problem in these scenarios was the failure to appreciate that software safety is not only a function of the individual code modules but also of the complete system. Hence, a module that was proven safe in one "context of use" (or environment – in this case a prior version of the machine) cannot be automatically assumed to be safe in a new context without extensive testing. There are bugs that do not manifest themselves in certain use cases that will appear when the software is stressed further.[7] As Leveson states:

Safety is a quality of the system in which the software is used; it is not a quality of the software itself. Rewriting the entire software in order to get a clean and simple design may be safer in many cases.

2.2   A Review of Relevant FDA Documents

In this section we will first review three important FDA documents: (1) the Quality System Regulations [69] (QSR or CFR 820 – Section 2.2.1); (2) the General Principles of Software Validation [70] (GPSV – Section 2.2.2); and (3) Software as a Medical Device (SaMD): Clinical Evaluation [92] (Section 2.2.3). The section concludes with a brief discussion of the FDA's regulatory process (Section 2.3).

These three relatively brief documents constitute primary source materials on our topic. Before proceeding, we need to define three key terms, namely **law, regulation, and guidance**. We follow here the descriptions provided by the FDA on its own website.

Law, at the national or federal level, in the United States is enacted by an act of Congress. Laws tend to be written at a high level, leaving the detailed implementation aspects to regulations that are issued by the appropriate government agency (in this case the FDA). In this particular case, the relevant legislation is the Federal Food, Drug, and Cosmetic Act (FD&C Act) of 1938 and subsequent amendments.

A regulatory agency, such as the FDA, develops **regulations** based on the laws, following the procedures required by the Administrative Procedure Act, another federal law, to issue regulations. This typically involves a process known as "notice and comment rulemaking" that allows for public input on a proposed regulation before the agency issues a final regulation. FDA regulations also have the force of law, but they are not part of the FD&C Act. The FDA regulations that govern medical devices can be found in Title 21 of the Code of Federal Regulations (CFR). This is the QSR document [69] that we previously referenced and which we will review extensively in Section 2.2.1.

The FDA follows the procedures required by its "Good Guidance Practice" regulation to issue **guidance** documents. While the content of these is not legally binding on the public or the FDA, they do describe the agency's current thinking on a regulatory issue and should not be taken lightly. The most relevant guidance documents for our purposes are the GPSV [70], which we discuss in some detail in Section 2.2.2, and the more recent document on the clinical evaluation of SaMD devices [92] (see Section 2.2.3). The guidance documents are often issued in draft form to allow for feedback (e.g. the new AI guidance document [82]) prior to finalization.

2.2.1   The Quality System Regulation – CFR 820

This is a relatively short document of about 58 pages of preamble material that describes the rationale behind the regulations, and a total of about 7 pages describing the regulations themselves, beginning on page 54 of the original document! These 7 pages are in the Code of Federal Regulations (CFR) Part 820 (which is why it is often referred to as CFR 820), titled *Quality System Regulation*, and provide very high-level instructions for how the process of designing medical devices (such that the FDA will clear/approve them) works.[8] It consists of 12 short subsections (labeled as subparts) as follows:

A. General Provisions (Scope, Definitions, Quality System)
B. Quality System Requirements (Management Responsibility, Quality Audit, Personnel)
C. Design Controls
D. Document Controls
E. Purchasing Controls
F. Identification and Traceability
G. Production and Process Controls (Production and Process Controls, Inspection, Measuring, and Test equipment, Process Validation)
H. Acceptance Activities (Receiving, In-process, and Finished Device Acceptance, Acceptance Status)
I. Nonconforming Product
J. Corrective and Preventive Action
K. Labeling and Packaging Control
L. Handling, Storage, Distribution, and Installation.

Given that these 12 sections fit into 7 pages, they generally provide a very short, high-level description of the material. In looking through this document, one is struck by several comments with inserted highlights in bold font, such as:

The requirements in this part govern the methods used in, and the facilities and controls used for, the design, manufacture, packaging, labeling, storage, installation, and servicing of all finished devices intended for human use. (820.1 Scope. a. Applicability)

Please note that the design of the process is governed by the actual regulations. How you design and build your software product is not simply an act of creative design, but an act of regulated (and, no doubt, creative) design.

The next interesting line (in the same section) states:

In this regulation, the term "where appropriate" is used several times. When a requirement is qualified by "where appropriate," it is deemed to be "appropriate" unless the manufacturer can document justification otherwise.

In plain English, this means that we (the FDA) get to decide what is appropriate, not you. This "where appropriate" formulation, however, also suggests a certain measure of regulatory flexibility on the part of the agency.

We will discuss how to create specifications, verification strategy, and validation strategy in Part III. Verification and validation are often confused or combined (you will see terms like "V&V" in some cases). Vogel [262, p. 85] has the following illustration that you may find useful:

It often has been said that verification activities ensure that the device was implemented correctly and validation activities ensure that the right device was developed.

Finally, as Vogel [262, p. 35] reminds us, validation in this context is as much about defect prevention as it is about defect detection and elimination. Design validation is an important component of the process. Time spent critically reviewing requirements and designs can often lead to finding defects before they occur and, thus, save the engineering effort "expended in implementing defective requirements or designs" (we develop the "wrong device"). Please keep in mind that the cost of dealing with defects rises exponentially as we move from concept to product. As a non-medical example, consider the recent case of the exploding batteries in the Samsung Note 7 phones [182]. The cost to Samsung of dealing with this via a recall was dramatically higher than it would have been had this been caught prior to mass sales (and even better, at a phase prior to mass manufacture).

Quality System (Subpart B) After the definitions, this second section deals with the design of a quality system. This describes first the responsibility of company management for the overall process, and then procedures for quality audits and personnel requirements and training. The regulatory requirements for a quality system are at a very high level. As Vogel summarizes, they basically suggest that one needs (1) an organizational system that supports compliance with the quality system; (2) formal procedures that define the system; (3) defined responsibilities within the organization; and (4) routine review and improvement of the system. Essentially, these say that an organization needs to take the creation of a quality system seriously and work "inside it" as opposed to "against it." See also our description of QMS in Chapter 4.

A key element of a good system is the concept of "review, revise, and approve" (RRA). Essentially, each process requires review (of the requirements, the design, the code, the tests, etc.), revision (if anything needs to be changed), and approval (a dated signature by a responsible official) that certifies the process. As with any regulatory-related issue, dates and signatures are important parts of the process. All such documents are archived in the design history file (DHF) of the device.

The document defines the DHF as the "compilation of records which describes the design history of a finished device."

Design Controls (Subpart C) This section discusses how to manage the design process appropriately. Here are some important quotes:

Each manufacturer . . . shall establish and maintain procedures to control the design of the device in order to ensure that specified design requirements are met.

Each manufacturer shall establish and maintain plans that describe or reference the design and development activities and define responsibility for implementation.

An illustration of the design controls process comes from the FDA's Quality Controls Guidance [88] and is shown in Figure 2.1. The development process begins with establishing the needs of the user. Once this is done, the needs are formalized/documented ("Design Input") as a set of system requirements that will guide the process. These requirements then drive the actual design process, which in our case is the process of designing our proposed software. At the end of the design process, we have our software design ("Design Output"). This design must be verified to ensure it meets the requirements; for software, verification also includes the regular software testing processes, performed during and after implementation. Finally, we have our medical device/software, which must be 'validated' to ensure that it meets the actual

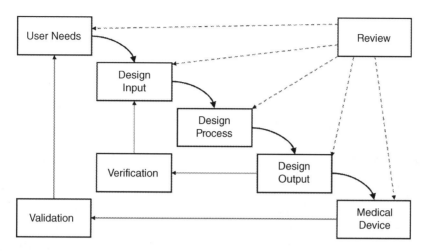

Figure 2.1 The FDA Waterfall model. This figure shows the Waterfall model repurposed for medical device development. The figure is adapted from figure 1 of the FDA Quality Controls Guidance [88], which in turn derives from documentation from the Medical Devices Bureau, Health Canada.

user needs (to check that our understanding of these needs was correct). Throughout the process, one must operate a system of formal reviews to ensure that each step was successfully completed, problems were identified, risks were managed, and evidence is assembled to prove the process was followed – that is, that the software was properly designed and constructed under "design controls."

More formally, the document defines some of these terms (Subpart A) as follows:

- Design input means the physical and performance requirements of a device that are used as a basis for device design. (IEC 62304 helpfully clarifies that the design inputs "are the interpretation of customer needs into formally documented medical device requirements" [119].)
- Design output means the results of a design effort at each design phase and at the end of the total design effort. The finished design output is the basis for the device master record. The total finished design output consists of the device, its packaging and labeling, and the device master record.
- Specification means any requirement with which a product, process, service, or other activity must conform.
- Validation means confirmation by examination and provision of objective evidence that the particular requirements for a specific intended use can be consistently fulfilled. (1) Process validation means establishing by objective evidence that a process consistently produces a result or product meeting its predetermined specifications. (2) Design validation means establishing by objective evidence that device specifications conform with user needs and intended use(s).
- Verification means confirmation by examination and provision of objective evidence that specified requirements have been fulfilled.

Finally, there is a comment that is repeated (with slight changes in vocabulary) at the end of many of these subsections (quoting from "Design Review") that reads:

The results of a design review, including identification of the design, the date, and the individual(s) performing the review, shall be documented in the design history file (the DHF).

The implication is that in a formal design process, we need formal (signed, dated, and properly archived) documents that verify that the process was followed.

Subparts D–O The next part, Subpart D, naturally then deals with how to keep formal records. Subparts E–O will not concern us here in this introduction, and are more appropriate for hardware than software.

A final comment on the QSR: The take-home lesson from reading through the QSR is that the medical software process calls for significant amounts of what some may derisively label as paperwork.[9] This is not the world of the ninja programmer, but rather the world of a serious engineering process not unlike that used in building a major bridge or an airplane. We are dealing with software that interfaces with devices (or is used by physicians) that treat real people, and our mistakes and bugs can have serious consequences. A Google search result for, say, "pizza restaurants near my location" is perfectly acceptable if 95 percent of the returned links are for pizza places. A medical device may not be acceptable, however, if 5 percent of the patients treated with it suffer serious injury (or worse).[10]

Please note the particular vocabulary used in the validation definition (above), including the words "objective evidence." Vogel comments (with respect to the GPSV, but the comment applies here too) that the FDA is a regulatory agency and that the choice of these words (as opposed to documentation, results, records, etc.) suggests that the authors of the regulations fully expect that "the evidence provided to establish the validated state of the device software could in fact be used as evidence in legal proceedings" [261].

The reader is encouraged to read this short document in its entirety.

2.2.2   The General Principles of Software Validation

This guidance document was issued by the FDA on January 11, 2002, and it supersedes a prior version dated June 9, 1997. This is an explanatory document and is meant to guide the software engineer toward best practices that will ensure that the end product meets FDA standards. The GPSV is divided into the following six sections:

1. Purpose;
2. Scope;
3. Context for Software Validation;
4. Principles of Software Validation;
5. Activities and Tasks;
6. Validation of Automated Process Equipment and Quality System Software.

Purpose (Section 1) This is short and to the point:

This guidance outlines general validation principles that the Food and Drug Administration (FDA) considers to be applicable to the validation of medical device software or the validation of software used to design, develop, or manufacture medical devices.

Scope (Section 2) This is also fairly short. We will review here a couple of key excerpts. First, it is clear that the FDA's idea of validation is broader than the one often used in pure software engineering:

The scope of this guidance is somewhat broader than the scope of validation in the strictest definition of that term. Planning, verification, testing, traceability, configuration management, and many other aspects of good software engineering discussed in this guidance are important activities that together help to support a final conclusion that software is validated.

Next comes a recommendation that the project is structured with explicit software life cycle management and risk management activities. This means that at any given time, the software team knows what phase it is in and does not arbitrarily switch from phase to phase. Per the GPSV:

This guidance recommends an integration of software life cycle management and risk management activities ... While this guidance does not recommend any specific life cycle model or any specific technique or method, it does recommend that software validation and verification activities be conducted throughout the entire software life cycle.

Finally, there is a critical comment on the use of external components (such as the operating system and other off-the-shelf libraries), which is the responsibility of the device maker to check:

Where the software is developed by someone other than the device manufacturer (*e.g.* off-the-shelf software), the software developer may not be directly responsible for compliance with FDA regulations. In that case, the party with regulatory responsibility (*i.e.* the device manufacturer) needs to assess the adequacy of the off-the-shelf software developer's activities and determine what additional efforts are needed to establish that the software is validated for the device manufacturer's intended use.

Later, in Section 2.4, there is an interesting set of statistics that makes the point that often things go wrong as part of software updates:

The FDA's analysis of 3140 medical device recalls conducted between 1992 and 1998 revealed that 242 of them (7.7%) are attributable to software failures. Of those software related recalls, 192 (or 79%) were caused by software defects that were introduced when changes were made to the software after its initial production and distribution.

Context for Software Validation (Section 3) This section begins with a statement to the effect that the FDA has received many questions as to how to ensure compliance with the QSR with respect to software validation. Their general response is that due to the great variety of medical devices and processes, it is not possible to give a single answer, but they suggest that a general application of broad concepts can be successfully used as guidance. Next, the section provides a set of definitions of key terms including requirement, specification, verification, and validation. These are expanded definitions of those found in the QSR itself. After the definitions, there is an interesting acknowledgment of the difficulties involved:

Software verification and validation are difficult because a developer cannot test forever, and it is hard to know how much evidence is enough. In large measure, software validation is a matter of developing a "level of confidence" that the device meets all requirements and user expectations for the software automated functions and features of the device ... The level of confidence, and, therefore, the level of software validation, verification, and testing effort needed, will vary depending upon the safety risk (hazard) posed by the automated functions of the device.

Hence, the effort that is needed as part of these tasks is proportional to the potential of our device/software to cause serious harm to the patient and/or user.

The next subsection discusses software development as part of the overall system design process. A key point is made that in the case of a medical device consisting of both software and hardware, software validation is part of the overall system validation. We will return to this topic in Chapter 15.

Differences between Hardware and Software Design Section 3 continues with a very interesting discussion (Section 3.3) titled *Software is Different from Hardware*, which focuses on issues that make software validation different from hardware validation. Some of the points made are as follows:[11]

- The vast majority of software problems are traceable to errors made during the design and development process, unlike hardware, whose quality is also dependent on the quality of the manufacturing process.
- Typically, testing alone cannot fully verify that software is complete and correct. In addition to testing, other verification techniques and a structured and documented development process should be combined to ensure a comprehensive validation approach.
- Unlike hardware, which wears out, software does not. However, updates in software can introduce defects.
- Unlike some hardware failures, software failures occur without advanced warning. This is because software can follow many paths (due to branching statements), some of which may not have been fully tested during validation. Also, seemingly insignificant changes in software code can create unexpected and very significant problems elsewhere in the software program.
- Given the high demand for software professionals, the software personnel who make maintenance changes to software may not have been involved in the original software development. Therefore, accurate and thorough documentation is essential.

The conclusion is worth quoting in full (emphasis is in the original document): **"For these and other reasons, software engineering needs an even greater level of managerial scrutiny and control than does hardware engineering."** Software is a critical, and often the most complex, part of the medical device. This is also becoming the case in other areas such as cars.

Section 3.4 discusses the benefits of software validation and points out that in addition to the direct goal of assuring quality and safety, proper validation can also reduce long-term costs by making it easier and less costly to maintain and update the software: "An established comprehensive software validation process helps to reduce the long-term cost of software by reducing the cost of validation for each subsequent release."

New releases may simply involve revalidating the same software to run on newer hardware or a newer version of the underlying operating system. One can then simply follow the same procedure as in the prior release to establish that the software still works as before.

The last part of Section 3 (Section 3.5) discusses design reviews. Please note, again, the emphasis on process as opposed to just results. The process via which the software is developed and validated (and the explicit suggestion for design reviews) is critical in establishing confidence that the software works. The document provides a short checklist for what the reviews should do (in part). During such reviews, answers to some key questions should be documented. These include:

- Have the appropriate tasks and expected results, outputs, or products been established for each software life cycle activity?
- Do the tasks and expected results, outputs, or products of each software life cycle activity:
  - comply with the requirements of other software life cycle activities in terms of correctness, completeness, consistency, and accuracy?
  - satisfy the standards, practices, and conventions of that activity?
  - establish a proper basis for initiating tasks for the next software life cycle activity?

The obvious take-home lesson is that the FDA expects software manufacturers to perform and document formal design reviews as part of the design, implementation, and validation of the software.

Principles of Software Validation (Section 4) This section highlights key principles for this process. It begins with the explicit statement that one needs a formal specification: "The software validation process cannot be completed without an established software requirements specification." **Many new startup companies are under the false impression that we first develop the software in some fashion and we will worry about the FDA when it comes to validating it.** This is emphatically rejected by this statement. Next, the document reemphasizes that the best validation strategy is defect prevention (much like in any other disease setting!) and that the process needs to focus on preventing defects rather than relying on testing to add quality into the system after it is written. We are reminded (and as anybody who has ever written regression tests will surely agree) that software testing "is very limited in its ability to surface all latent defects in the code." Software is too complex to check every possible branching path and parameter setting. Then, it states in bold font:

Software testing is a necessary activity. However, in most cases software testing by itself is not sufficient to establish confidence that the software is fit for its intended use.

Next, we move onto a statement that this takes time and effort and then find statements (more on this in Section 2.3 of this chapter) about life cycles, plans, and procedures, which we will revisit in Chapter 15.

Section 4.7 discusses the important topic of how to revalidate software after a change. It reminds us (and, no doubt, the authors have the Therac-25 in mind here) that even "a seemingly small local change may have a significant global system impact." Software is similar to nonlinear dynamical systems. We are used to linear systems in everyday life, where, typically, small changes in the input produce small changes in the output. In some nonlinear systems, by contrast,[12] even minor changes in the inputs can lead to massive changes in the output. Hence, the GPSV (again in bold) states that whenever "software is changed, a validation analysis should be conducted not just for validation of the individual change but also to determine the extent and impact of that change on the entire software system."

Section 4.8 discusses validation coverage and makes the point that the degree of this should relate to the software's complexity and risk, not to the size of the company or resource constraints! Section 4.9 discusses the key concept of independence of review and reminds us that self-validation is extremely difficult. The person performing the validation should be independent from the developer (hence, avoiding the common developer aphorism "it has always worked for me!"), going as far as to suggest that sometimes a third-party independent firm might be contracted to perform this task.

Section 4.10, which is titled *Flexibility and Responsibility,* makes two interesting points. The first is that the application of the principles in this document may be quite different from one instance to another. The FDA provides guidance but does not mandate specific procedures at the detail level. However, we are reminded at the end that "regardless of the distribution of tasks, contractual relations, source of components, or the development environment, the device manufacturer or specification developer retains ultimate responsibility for ensuring that the software is validated."

Activities and Tasks (Section 5) This relatively long section (which is one-third of the total GPSV) provides a detailed description of many activities and tasks that are appropriate for validation. We will rely on many of these descriptions in Part III of this book. The activities and tasks are divided into subsections titled:

- Quality Planning;
- Requirements;
- Designs;
- Construction or Coding;
- Testing by the Software Developer;
- User Site Testing;
- Maintenance and Software Changes.

We will highlight here aspects of "Designs" (Section 5.2.3), which relates to what is termed *human factors*. The focus of this particular section is to avoid user error. As the document states, one of the most common and serious problems that the FDA finds is caused by designs that are either too complex or operate in ways that users find counterintuitive. A related issue is that similar devices made by different vendors may have different user interfaces, causing errors by the user as they switch between devices. The recommendation is to use human factors engineering techniques, which should be used throughout the design and the development process from the requirements phase to the validation phase – we discuss this topic in more detail in Section 13.5. In addition to preventing error, human factors considers the design of the product to allow for software designed to best meet the users' needs in a way that is most facile to use. Particular importance is placed on having actual users (e.g. having surgeons participate in the testing of software for surgical planning) participate in the testing as part of the application of such methodologies. The FDA has an additional guidance document that discusses human factors in more detail [89]. See also the international standard IEC 62366 [122, 124].

Validation of Automated Process Equipment and Quality System Software (Section 6) This last section includes a useful discussion on "off-the-shelf software" that is worth quoting in full:

For some off-the-shelf software development tools, such as software compilers, linkers, editors, and operating systems, exhaustive black-box testing by the device manufacturer may be impractical. [See Section 9.5 for a definition of black-box testing – this is essentially testing where the tester has no knowledge of the internal structure of the software being tested.] Without such testing – a key element of the validation effort – it may not be possible to validate these software tools. However, their proper operation may be satisfactorily inferred by other means. For example, compilers are frequently certified by independent third-party testing, and commercial software products may have "bug lists," system requirements and other operational information available from the vendor that can be compared to the device manufacturer's intended use to help focus the "black-box" testing effort. Off-the-shelf operating systems need not be validated as a separate program. However, system-level validation testing of the application software should address all the operating system services used, including maximum loading conditions, file operations, handling of system error conditions, and memory constraints that may be applicable to the intended use of the application program.

In practical terms, a modern software application runs on top of a standard operating system. In addition, it may use, potentially, many software libraries for parts of its operation. It also depends on development tools for its creation, compilation, and testing. Ultimately, it also depends on microcode and circuits in the computer hardware itself. These, while not part of our software, can also be a source of potential problems, and they need to examined for issues too, though not at the same level as our own code.

The GPSV is worth reading in full. It is about 42 pages long, and it contains much valuable information that the aspiring medical software developer should be aware of.

2.2.3   The Software as Medical Device (SaMD): Clinical Evaluation Document

This newer guidance document addresses significant changes that have occurred since 2002, when the GPSV [70] appeared. This document was produced by the IMDRF and reissued by the FDA as guidance. The IMDRF (as stated on its website) was conceived in February 2011 as a forum to discuss future directions in medical device regulatory harmonization. It is a voluntary group of medical device regulators from around the world who have come together to build on the strong foundational work of the Global Harmonization Task Force on Medical Devices (GHTF) and to accelerate international medical device regulatory harmonization and convergence.

In particular, as stated in the document, the "IMDRF has acknowledged that software is an increasingly critical area of healthcare product development," and, in particular, the use of standalone software (not attached to specific medical devices) as a healthcare delivery vehicle. This document builds upon three prior documents produced by the IMDRF. These are (1) "Software as a Medical Device (SaMD): Key Definitions" [125]); (2) "software as Medical Device: Possible Framework for Risk Categorization and Corresponding Considerations"s [126]; and (3) "Software as Medical Device (SaMD): Application of Quality Management System" [127]. We provide a short summary in this section and review some of these documents in more detail in Chapters 4 (QMS), 5 (risk), and 15 (validation). An additional document from the IMDRF discusses cybersecurity [129]. We briefly review this in Section 3.5.

The "Prerequisite" Documents

Definitions The IMDRF [125] defines the term "software as a medical device" as "software intended to be used for one or more medical purposes that perform these purposes without being part of a hardware medical device." The document then provides the following notes:

- SaMD is a medical device and includes in-vitro diagnostic (IVD) medical devices.
- SaMD is capable of running on general purpose (non-medical purpose) computing platforms.
- "Without being part of" means software not necessary for a hardware medical device to achieve its intended medical purpose.
- Software does not meet the definition of SaMD if its intended purpose is to drive a hardware medical device.
- SaMD may be used in combination (e.g. as a module) with other products, including medical devices.
- SaMD may be interfaced with other medical devices, including hardware medical devices and other SaMD software, as well as general purpose software.
- Mobile apps that meet the definition above are considered SaMD.

Please note that mobile apps can be considered SaMD, and are subject to regulation. The FDA has an interesting guidance document on this topic [80].

SaMD Characterization The second document in the series (*Software as Medical Device: Possible Framework for Risk categorization and Corresponding Considerations* [126]) then provides "recipes" for the categorization of the software, depending on the risks it poses (more in the next section). The document proposes four categories (I, II, III or IV) based on "the levels of impact on the patient or public health where accurate information provided by the SaMD to treat or diagnose, drive or inform clinical management

Table 2.1 IMDRF Risk Categorization. A summary of how the IMDRF categorizes software in terms of its use and its potential impact on a patient. Reproduced from Section 7.2 of the IMDRF Risk Categorization Document [126].

State of healthcare situation or condition	Significance of information provided by SaMD to healthcare decision		
	Treat or diagnose	Drive clinical management	Inform clinical management
Critical	IV	III	II
Serious	III	II	I
Non-serious	II	I	I

Table 2.2 FDA safety classification table and premarket review process. A summary of the safety classification for medical devices (Class I, II, or III) and the regulatory controls and premarket review processes that apply to each class.

Class	Risk	Regulatory controls	Review process
I	Low to moderate	General Controls	Most are exempt
II	Moderate to high	General Controls Special Controls	Premarket Notification 510 (k), or De Novo
III	High	General controls Premarket Approval	Premarket Approval (PMA)

is vital to avoid death, long-term disability or other serious deterioration of health, mitigating public health" [126]. Here, Category I has the lowest significance/risk and Category IV, the highest. The document nicely summarizes its classification strategy in the table reproduced here as Table 2.1. In general, this is similar to the FDA device classification table presented in Table 2.2.[13] The risk level increases as we move from a direct impact on clinical decision making (from "Inform" to "Treat or Diagnose") and the state of the patient ("non-serious" to "critical"). This is fairly standard risk management methodology (see Chapter 5) and corresponds to risk, defined as the combination of the severity of harm and the probability of harm that can be caused by any operation/event. The document provides many useful examples that can help any particular company/designer assess the risk categorization of their proposed software.

The SaMD Clinical Evaluation Document

Clinical Evaluation Summary The clinical evaluation of SaMD[14] is divided into three categories (see also Table 15.1). The first is the demonstration of clinical association. This answers the question "How much meaning and confidence can be assigned to the clinical significance of the SaMD's output?" The second aspect of this evaluation

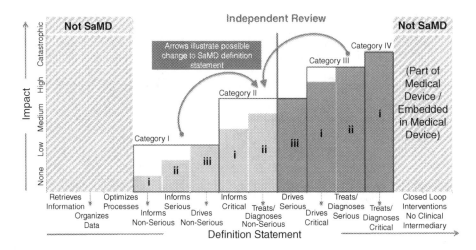

Figure 2.2 Risk level categorization for SaMD. A summary of the methodology for categorising SaMD. Reproduced from figure 13 of the FDA SaMD Clinical Evaluation document [92].

is analytical validation. Here, the goal is to confirm and provide "objective evidence" that the SaMD meets key technical requirements (software is correctly constructed, meets specifications, and conforms to user needs). The final aspect of the clinical evaluation is clinical validation. Here, the goal is to measure the ability of SaMD to yield clinically meaningful (i.e. one that has positive impact on the health of an individual or population) output associated with a disease process. We discuss this topic in more detail in Section 15.1.1 as part of our detailed discussion of validation.

Independence of Review Section 8 of the SaMD [92] document stresses the importance of independent review of an SaMD's clinical evaluation. In general, as the risk level of the SaMD increases, the recommendation (which is true of all software) is that the clinical review is performed by an independent entity. Figure 2.2 provides a useful categorization of what is and what is not SaMD, the category risk levels, and the types of categories for which independent review is more important.

2.3 The FDA Regulatory Process

Device Classification Bringing a medical device to market in the United States can be a complex process. To understand the level of complexity your company's device will face, you have to define both the intended use and the indications of use for the device.[15] Once this has been done, it is possible to classify your device according to the FDA's risk-based system of device classification. Devices are divided into classes, as shown in Table 2.2.

Class I devices are considered to have low to moderate risk and require "General Controls" [74] – which are the "basic provisions" and legal requirements for all classes of devices.[16] Most Class I devices do not require premarket review. Examples include wheelchairs and surgical gloves.

Class II devices are those that have moderate to high risk and are subject to both general and special controls. Special controls are usually device-specific and include meeting certain performance standards[17] and having procedures for postmarket surveillance [74]. Class II devices are cleared either through the Premarket Notification 510(k) [86] process or the De Novo process [77]. In the 510(k) process, the FDA determines that the device is substantially equivalent and has a similar safety profile to a previously commercialized product. The device from which equivalence is drawn is commonly known as the "predicate." Additionally, the newer De Novo pathway is used to provide novel devices with Type I or Type II classification based on reasonable assurance of safety and effectiveness for the intended use in the absence of predicates. Examples of Class II devices include medical imaging systems and most AI/ML software.[18]

Class III devices are those that are considered high risk, defined as "supports or sustains human life, is of substantial importance in preventing impairment of human health, or presents a potential, unreasonable risk of illness or injury" such as cardiac pacemakers or novel imaging devices (e.g. recently digital breast tomosynthesis devices). These are reviewed using the PMA [81], where one must "demonstrate reasonable assurance of safety and effectiveness." The notification pathway under 510(k) clearance may be considerably faster and less expensive than the PMA route. Therefore, manufacturers and product developers often closely examine the possibility of following the clearance pathway, which may turn on the likelihood of finding an appropriate point of comparison, or predicate device, agreeable to the FDA.

A recent survey by Benjamens et al. [19] listed 29 AI/ML-based medical technologies that have received FDA approval or clearance. Of these, 5 used the De Novo pathway, 24 used the 510(k) premarket notification process, and only 1 (a product used to predict blood glucose changes) used the PMA process.

Documentation Required The FDA document *Guidance for the Content of Premarket Submissions for Software Contained in Medical Devices* [71] provides the following summary description of the documentation required (for software) in a submission:

- describe the design of your device;
- document how your design was implemented;
- demonstrate how the device produced by your design implementation was tested;
- show that you identified hazards appropriately and managed risks effectively; and
- provide traceability to link together design, implementation, testing, and risk management.

We will discuss these topics in Part III. For example, the design of the device (software in our case) maps to the System Requirements Specification, a simplified template for which is presented in Chapter 12. The level of detail required for each of these components depends on the device classification.

It is worth noting that the FDA has procedures for "pre-submission" queries that allow manufacturers to obtain early input/feedback from the FDA prior to an intended premarket submission. While this program is entirely voluntary on the part of the manufacturer, it can be particularly valuable, especially for the design of clinical trials.

The EU Regulatory Clearance Process The process in Europe is somewhat different. In the European Economic Area, a CE Mark[19] is required to bring a medical device to market. By contrast to the United States, where the FDA is both the regulator and the reviewing authority, in the EU these functions are divided among three different bodies. The first is the European Commission, which sets high-level transnational policy. Next, we have the competent authorities at the national level (i.e. a different body for every member state of the EU), which monitor the market in their own country. Finally, we have notified bodies, which are private entities that are authorized (and audited) by the government authorities to perform medical device reviews and to issue certificates of conformance (CE Mark) for specific types of devices. In the EU, therefore, the regulatory review is performed by these private entities and not by a government agency.

2.4 Regulatory Documents from Outside the United States

Other international entities (countries or zones in the case of the European Economic Area[20]) have similar regulations and processes to that of the United States. There is a movement toward international harmonization, as evidenced by the work of the IMDRF. As of early 2021, the following countries had agencies that were members of the IMDRF: Australia, Brazil, Canada, China, EU countries, Japan, Russia, Singapore, South Korea, and the United States. What follows is a brief survey and references to some important regulatory documents produced by these countries that should be considered when designing medical software.

The core EU regulation is Medical Devices Regulation 2017/745 [64] that took effect in May 2021 – this is commonly referred to as the MDR. The MDR is a significant revision of previous EU regulations (the Medical Device Directive, or MDD, of 1993), partially in response to some high-profile device scandals (e.g. the PIP breast implant scandal [165]). There is also a recent guidance document on the application of aspects of the MDR to software [62]. While the European Medicines Agency is the regulatory agency for medical devices in the EU, there are also national institutions that issue useful guidance and standards documents as the German Institute for Standardization. We use the AI guidance document produced by this institute [53] as a key reference in multiple sections.

In China, the process is overseen by the National Medical Products Administration (NMPA) – see, for example, the *Rules of Classification of Medical Devices* [188]. Unfortunately, most of these documents are not available in English.[21] In Japan, the regulatory process is overseen by the Pharmaceuticals and Medical Devices Agency

(PMDA) under the PMD Act [102]. India had no medical device regulations until recently (around 2005). The latest medical device rules were issued in 2017 [120]. The South Korean regulator (the Ministry of Food and Drug Safety) has a comprehensive set of regulations that can be found on their website [214].

There is a good description of the regulatory constraints for human factors in a guidance document by the UK Medicines & Healthcare Products Regulatory Agency (MHRA), titled *Human Factors and Usability Engineering – Guidance for Medical Devices Including Drug–Device Combination Products* [173]. A second interesting MHRA document is *Guidance: Medical Device Stand-Alone Software Including Apps (Including IVDMDs)* [174]. This includes a figure – which we reproduce as Figure 13.1 – that demonstrates how to label cleared mobile apps.[22]

Two other very readable documents, available in English, are the *Software as a Medical Device (SaMD)* [109] guidance document from Health Canada and the *Regulatory Guidelines for Software Medical Devices – A Life Cycle Approach* [234] document from the Health Sciences Authority of Singapore. This last document is both relatively brief (32 pages) and also very comprehensive in at least highlighting all aspects of medical software work, from quality management systems (see Chapter 4), to cybersecurity (see Section 3.5), to issues relating to AI.

One more document worth referencing here is a recent literature review from the Therapeutic Goods Administration of Australia, titled *Actual and Potential Harm Caused by Medical Software* [252]. This reviews a number of publications that have documented medical software failures. In the conclusion, the paper notes that "The tendency for journals to publish positive findings and for negative reports to struggle to be published means that there is only limited research into the safety of software when used in a medical setting and that studies indicating performance failures for apps are often not published [24]."

Finally, some countries (e.g. Israel) permit the sale of medical devices provided the manufacturer can provide approval in some other countries, such as evidence of FDA 510(k) clearance (USA) or CE Mark (European Union).

2.5　Implications of Regulatory Requirements for Software Design

An Explicit Plan is Needed As the regulatory documents make clear, the structure and the environment in which the software development takes place are critical. We must be able to establish procedures and plans that demonstrate the overall software development process is controlled and follows an actual plan as opposed to ad-hoc programming. In particular, to satisfy the need for "design controls" we must:[23]

1. establish before any actual programming what we are going to do – **the plan**;
2. follow **the plan**; and
3. be able to provide documentary evidence that **the plan** was followed.

The process of creating FDA-"clearable" software begins at the planning stage. It does not begin (and we will repeat this point many times) at the validation phase.

Everything Must be Documented This is spelled out as the Design History File (DHF) in the Quality System Regulation (QSR), which defines this as "*a compilation of records which describes the design history of a finished device*" [69]. All design documents must be formally reviewed and approved (signatures and dates). This is needed as part of being able to provide documentary evidence. As the AAMI TIR 45 document states: "a common principle in the medical device software world is 'If it isn't documented, it didn't happen' " [1].

One Must Use an Explicit Software Life Cycle The FDA does not mandate a specific life cycle model (e.g. Waterfall or Agile – see Section 9.2), only that a life cycle is used. The reason for this is that a life cycle–based process (with a formal review at the end of each phase of the life cycle) builds confidence in the software ("ordered kitchen"). To quote the GPSV:

Software validation takes place within the environment of an established software life cycle. The software life cycle contains software engineering tasks and documentation necessary to support the software validation effort. In addition, the software life cycle contains specific verification and validation tasks that are appropriate for the intended use of the software. This guidance does not recommend any particular life cycle models – only that they should be selected and used for a software development project.

Essentially the FDA is telling us you can use any software cycle model you want, but you must use one! The alternative is uncontrolled software development, which does not inspire confidence in the final product.

As an example, consider the Waterfall life cycle. This consists of:

1. system requirements;
2. software design;
3. software implementation;
4. testing;
5. maintenance.

While the Waterfall model is simple and intuitive, it has been criticized for failing to reflect the realities of developing complex software. For example, at the design phase, we may discover that the requirements are incomplete and need to go back and revise the requirements. What following a life cycle model implies is that once we discover this problem, prior to continuing to work on the software, one should stop review and revise the specifications to reflect on the new discovery and then proceed to working on actual code. In this way, the documentation always reflects the plan, and a review of the software should reveal that we are always following a plan; hence, we are operating "under control" and not in an ad-hoc fashion. There are other life cycles – see Section 9.2.

Complexity The GPSV (as quoted in the previous section) notes that the existence of branching points (e.g. if -- then -- else statements) allow the software to follow many paths, which make testing and validation complicated. (An example of the effect of branching points on testing is given in Section 9.5.1.) This is particularly evident compared to hardware development, where a device may only operate in a few, easily enumerable modes. Software is infinitely flexible, and the price of such flexibility is often the inability to fully test the software. A recommendation here is to eliminate (or better yet, not introduce) unnecessary complexity in the design. The software should be designed with **the smallest possible feature set that will do the job**. The software designer needs to be able to say "no" to seemingly innocent user requests and needs unless these are absolutely necessary.

A related topic is that the supplier of medical software is liable for any issues that arise from its use (unless this is due to misuse). Since any addition to a piece of software adds potential bugs, one simple rule of thumb in medical software is to restrict the functionality of the code to the minimum required to meet users' needs. While this is good software engineering practice in general, in this domain it is also necessary to reduce liability risk. Simpler software is easier to test.

Longevity The other issue worth noting is that medical software often has a long lifetime, much longer than typical consumer-oriented software. Once one has to go through the regulatory process (which can be an expensive proposition), the incentive is to keep the software alive for as long as possible. This longevity imposes constraints on the selection of the underlying operating system (to ensure availability/backwards compatibility years down the road), and external dependencies (such as, for example, libraries for 3D graphics). In general, one should avoid using bleeding-edge libraries, but rather select ones that have stood the test of time and are likely to be around and maintained well into the future.

2.6   The Future: Digital Health, AI, Self-Driving Cars?

2.6.1   FDA: Digital Health and AI

In the past 2–3 years, the FDA has issued some very interesting new guidance documents that relate to the topics of digital health [90] and the use of AI and ML techniques [82, 91]. In addition, the FDA has recently introduced a remarkable pilot digital health software pre-certification pilot program [85] – see also Chapter 4 – where the review emphasis is shifting from reviewing the device (SaMD) to reviewing the manufacturer. This is an attempt by the agency to address issues created by the explosion in the application of deep learning algorithms in medicine [19].

The emergence of deep learning algorithms has been a source of great excitement. Such algorithms are trained using large amounts of data and then are applied to generate predictions for new data. The concept is not new; ML tools have been used for mammography and lung cancer evaluation for years under the general name of computer-aided diagnosis. What is new here is that the complexity of the algorithms (as measured in numbers of parameters that need to be learned) is many orders of magnitude larger in deep learning applications vs. conventional ML/pattern recognition tools. The FDA, in a recent guidance document [82], goes as far as to describe procedures for clearing devices in which the AI models are changed to improve performance in a way that does not require additional FDA review for each change by providing a "predetermined change control plan." This plan has two components. First, we have the types of anticipated modifications – this is labeled as the SaMD Pre-Specifications (SPS). Second, we have the associated methodology to "implement those changes in a controlled manner that manages risks to patients" – this is the Algorithm Change Protocol (ACP). The FDA will (this is proposed guidance), in this scenario, review and approve the change control plan, and the manufacturer can use these procedures to improve the algorithm without further review.

2.6.2   Asian and European Guidance

Other regulatory agencies such as the NMPA (China) are issuing their own guidelines [189] on AI-aided software. There is also an interesting report on this topic published by the PMDA of Japan [31]. In addition, there is a brief but very readable section on AI in the latest guidelines from Singapore's Health Sciences Authority [234]. Finally, the German Institute for Standardization (DIN) has also issued a recent document on life cycle processes for AI methods [53]. We will discuss these guidelines in more detail in Sections 9.3 and 15.1.2.

2.6.3   A Detailed Look at the Singapore Guidance

The guidance from Singapore's Health Sciences Authority [234] has a very nice section (Section 8 of this document) describing the regulatory requirements (in Singapore) for what they term AI-MD devices (i.e. medical devices that use AI technology). They note that these are extra on top of what is normally required for other medical devices. In particular, they request additional information on the following:

1. Dataset – in particular a description of the source, the size, the labeling and annotation, and a rationale for the choice.
2. AI model – the actual model used and a justification for its selection.
3. Performance and clinical evaluation – details on the performance and information on failure cases and other limitations.
4. Deployment – in particular a description of any human intervention (both timing and extent) and issues with retraining the model.
5. Continuous learning capabilities – particularly issues to ensure that the new data conforms to the standards of the original dataset, that no biases are introduced from new data, that there is a safety mechanism in place to detect anomalies, and the ability to roll-back to a previous version.

This is a good summary of the new issues that the use of AI/ML techniques present in medical software and the steps that regulators are taking to ensure that these medical devices are both safe and effective.

2.6.4 Non-Medical Areas: Self-Driving Cars?

The regulations and international standards for medical software can be profitably read by software engineers in other industries that involve mission-critical or complex software development. The same deep learning techniques that are being applied in medical applications are also finding their way into other applications, such as self-driving car navigation.

For both medical and non-medical products, if we follow the FDA triptych of efficacy, safety, and security as our evaluation standard, we may discover that while demonstrating efficacy may be reasonably straight-forward, safety and security pose challenges. In terms of safety, it is unclear how much testing is required to establish beyond reasonable doubt that these algorithms will be reliable in the real world. It is hard to even begin to catalog all the possible conditions. Security might be an even greater concern, given the development of adversarial attack methods for neural networks [6, 199, 244]. When the authors of the GPSV raised a concern about "branching statements" in the year 2002 (Section 9.5.1), they probably did not imagine the degree of complexity of the software that they would need to evaluate as these types of algorithms become more popular.

Increasingly, many manufacturers, such as car makers, are gradually becoming fully fledged software companies. The comment in the GPSV that "for these and other reasons, software engineering needs an even greater level of managerial scrutiny and control than does hardware engineering" [70] is just as applicable in this domain. One recent example is the issues that Volkswagen and other traditional car manufacturers are facing as they try to compete with Tesla. As a recent *Wall Street Journal* article points out in terms of some of the obstacles traditional car companies face: "What they didn't consider: Electric vehicles are more about software than hardware. And producing exquisitely engineered gas-powered cars doesn't translate into coding savvy" [21].

2.7 Conclusions

The take-home point of this chapter is that the regulatory process affects every single aspect of medical software throughout its life cycle, from design to release. Many new startup companies operate under the false impression that they can first develop the software in some fashion, and then worry about the FDA when it comes to validating it. This is emphatically rejected by the guidance documents that govern the entire process, beginning with the initial software design. Hence, a careful reading of the guidelines and related documents is a prerequisite for starting on this journey, as opposed to something one visits at the time of submitting an application for regulatory review. This principle applies equally to traditional software and new AI/ML-based tools.

2.8   Summary

- The regulatory process used by the FDA has evolved over the past 100 years to account for changing situations and in response to major incidents. For medical software, the origin story is the case of the Therac-25, which led the agency to better appreciate the potential dangers caused by errors in software as opposed to just hardware.
- There are a number of key regulatory documents produced both by the FDA and the IMDRF that govern this process, which we reviewed in some detail in this chapter. These are well worth reading in their entirety.
- The risk classification of a medical software device has a direct impact on both how the tool should be designed and validated and the rigor of the regulatory review process. In the United States, the most common pathways are the 510(k) clearance process (for those devices that are equivalent to something already approved), the De Novo process for new software that is of low to medium risk, and the PMA process for high-risk software.
- The process in most regulatory zones is largely similar to that in the United States, though there are some differences between them that one should be aware of when developing for these markets.
- The design, implementation, and validation of medical software is critically impacted by the regulatory structure. In particular, there are constraints on both the organization (quality management systems) and the process (use of a software life cycle, documentation, risk management) that the software engineer must be aware of when embarking on this process.
- The growth in the use of AI/ML techniques is prompting regulatory agencies to create new guidelines for software that uses these, so as to address the unique problems (and opportunities) in this area.

RECOMMENDED READING AND RESOURCES

For an overview of the FDA's regulatory process (this covers devices, but also drugs and biologics), see:

> E. Whitmore. *Development of FDA-regulated Medical Products: A Translational Approach.* ASQ Quality Press, 2nd ed., 2012.

The following book covers much useful material and should be read by those managing industry medical software projects:

D.A. Vogel. *Medical Device Software Verification, Validation, and Compliance*. Artech House, 2011.

The two classic FDA documents are the QSR and the GPSV. We reviewed these in Sections 2.2.1 and 2.2.2. Both of these are relatively short and well worth reading in their entirety:

FDA. General principles of software validation; final guidance for industry and FDA staff, January 11, 2002.

FDA. Medical devices: current good manufacturing practice final rule – quality system regulation. *Federal Register*, 61(195), 1996.

Probably the best example of a medical software regulatory document that covers the entire area is the following regulatory document from Singapore. This is both relatively brief (32 pages) and also very comprehensive in at least highlighting all aspects of medical software work from quality management systems (see Chapter 4), to cybersecurity (see Section 3.5), to issues relating to AI:

Singapore Health Sciences Authority (HSA). Regulatory guidelines for software medical devices – a lifecycle approach, December 2019.

For issues related to the use of AI/ML and medical software, the following three documents are worth reading:

FDA. Proposed regulatory framework for modifications to artificial intelligence/machine learning (AI/ML)-based software as a medical device (SaMD). Discussion Paper and request for feedback, April 2, 2019.

NMPA. Technical guidelines on AI-aided software, June 2019. (We used (and verified) an unofficial translation made available through the website of the consulting company China Med Device.)

A. Esteva, A. Robicquet, B. Ramsundar, et al. A guide to deep learning in healthcare. *Nat Med*, 25(1):24–29, 2019.

As an example of a discussion of the issues of using AI/ML techniques in non-medical areas, see this description of issues in automotive software by Salay and Czarnecki:

R. Salay and K. Czarnecki. Using machine learning safely in automotive software: an assessment and adaption of software process requirements in ISO 26262. arXiv:1808.01614, 2018.

NOTES

1. This is what the device is intended for. The full legal definition of this is complex and beyond the scope of a textbook. Under FDA regulations, the term "intended use" relates to the objective intent of the medical product manufacturer, and determining this may also involve, for example, reviewing claims made in advertising the product.

2. See also Appendices A and B of the FDA guidance document *Policy for Device Software Functions and Mobile Medical Applications* [80].

3. As an aside, some devices also go through lab, animal, pre-clinical, and clinical studies. This depends on the device and its intended use.

4. This section follows the discussion in the article "FDA's origin" [243] by John P. Swann, which appears on the FDA's website and which is in turn adapted from George Kurian's (ed.), "*A Historical Guide to the U.S. Government*" [150].

5. This builds on the original paper by N.G. Leveson and C.S. Turner [156]. We discuss this topic further in vignette presented in Chapter 17.

6. At least to somebody who has had some traditional software engineering training.

7. Moving from single-threaded to multi-threaded design is a particularly common source of new errors.

8. The QSR is in the process (beginning in 2018) of being harmonized with ISO 13485 [132]. This will allow for the harmonization of the quality system requirements in the United States with those in other regulatory zones, such as the EU.

9. A reviewer of this book commented that this paperwork burden unfortunately discourages creative programmers from working on regulated projects. He suggested that it is worth pointing out that creating a working prototype is often the best kind of design specification (or at least the best preparation work for a specification) because it addresses all the risk factors. Naturally this code cannot be reused as is, but one can use the prototype to develop the features and corresponding tests for the real system. Some of this code can then be migrated, piecemeal, from the prototype to the product with appropriate reviews and tests.

10. This depends on a risk–benefit analysis. If the severity of the patient's condition is high, then this (or even a higher) failure rate may be acceptable.

11. These bullet points are a mix of actual quotes and our paraphrasing (so as to abbreviate) from the GPSV.

12. Consider the well-known butterfly effect (see the book by Edward Lorenz [161]) which explains the difficulties in weather forecasting that persist to this day. See also the discussion in Littlewood and Strigini [158].

13. There are other schemes that use either three or four classes but share the same principles of moving from less serious to increasingly more serious risk of harm to a patient or caregiver. The international standard IEC 62304 [119] uses a three-class (A, B, or C) safety characterization for software. The NMPA (China) [188] also uses a three-class characterization (I, II, or III). The MDR (EU) [64] mandates a four-class categorization (I, IIa, IIb, or III).

14. Our description is heavily influenced by a presentation that Dr. Nicholas Petrick of the FDA gave at Yale – he acknowledged the help of Dr. Berkman Sahiner in preparing this.

15. As the FDA clarifies: "Indications for use for a device include a general description of the disease or condition the device will diagnose, treat, prevent, cure, or mitigate, including a description of the patient population for which the device is intended. Any differences related to sex, race/ethnicity, etc. should be included in the labeling" [73].

16. The Medical Device Amendments (1976) state that General Controls "include provisions that relate to adulteration; misbranding; device registration and listing; premarket notification; banned devices; notification, including repair, replacement, or refund; records and reports; restricted devices; and good manufacturing practices."

17. In the case of a 510(k) application these will be related to what was used to evaluate the predicate device.

18. Interestingly, in the new EU regulations that took effect in May 2021, the predicate-based clearance process is minimized and the equivalent of the De Novo process is preferred.

19. CE stands for "Conformitè Europëenne," which is translated as "European Conformity."

20. The European Economic Area consists of the member states of the EU plus Iceland, Lichtenstein, and Norway.

21. Unofficial translations of some of these documents can be found on the website of the consulting company China Med Device [30].

22. The MHRA is acquiring increased responsibilities as the process of the UK leaving the EU is completed. As stated in the guidance: "Please note that CE marked devices will continue to be recognised on the Great Britain market until 30 June 2023" [174].

23. We paraphrase from a presentation by Carl Wyrva titled *Product Software Development and FDA Regulations Software Development Practices and FDA Compliance* that was presented at the IEEE Orange County Computer Society in 2006.

3 Operating within a Healthcare System

INTRODUCTION

This chapter provides background on the constraints imposed on software by the need to operate within a healthcare environment. We first present an overview of the environment and the constraints under which our software must operate (Section 3.1). A description of the current healthcare environment (Section 3.2) follows. We begin with a description of the complex system currently in place in the United States, and then discuss how healthcare operates in the rest of the world. The next section discusses clinical information technology (Section 3.3), with an emphasis on electronic health records (EHR) and imaging databases (PACS). Following this, we review issues related to data privacy (HIPAA and GDPR) in Section 3.4. We conclude this chapter with a discussion of cybersecurity (Section 3.5). This is one of the primary sources of security concerns in medical software and a topic of increasing importance in today's connected world.

In addition to providing background material, this chapter addresses some of the questions that will need to be answered in our system requirements document – see Chapter 12. In particular, in Section 3.2 we describe the potential users for our software. Section 3.3 describes the information technology environment in which our software will live, which will form part of the use environment for our software. These topics will inevitably find their way into the non-functional requirements and risk management for our software.

3.1   Software as Part of an Ecosystem

John Donne's famous poem "No man is an island," written nearly 400 years ago, begins with the lines:

No man is an island entire of itself; every man is a piece of the continent, a part of the main . . .

We can paraphrase this poem to capture the essence of most medical software. "No medical software is an island entire of itself; every piece of software is a piece of a workflow; a part of the main (healthcare system)."

Modern healthcare systems are incredibly complex. There are many stakeholders involved in the provision of patient care, ranging from physicians to data-backup services. Figure 3.1 shows only a partial list of the stakeholders involved in the US

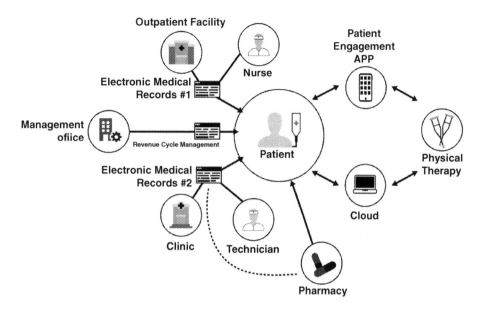

Figure 3.1 The Healthcare System in the United States. This is complex in terms of separating the users of medical software (doctors, nurses, patients, for example) from the economic buyer (e.g. outpatient facility). As shown here, when we try to make healthcare information more "patient-centric," it still involves communication between the patient app with multiple electronic medical records, depending on how many patient care facilities a person visits. This is a gross oversimplification as there can be multiple provider groups and facilities involved, but is intended to demonstrate the multi-directional flow of patient information. Figure by Ellie Gabriel.

healthcare system. Understanding who the stakeholders are and their needs, requirements, and regulatory constraints is a critical component of designing useful medical software.

Integrating our software with the rest of the healthcare environment is one of the central challenges in developing medical software in the modern era. Our software is not an island, and we must build bridges to other relevant software (and potentially hardware) systems in the healthcare environment. This will allow us to produce useful tools that become valuable parts of providing quality care to a patient. Before we build bridges, it is important to know what, to continue with the analogy, the remainder of the continent looks like and, in particular, what the major components of this continent are. Moreover, the healthcare environment is a constantly shifting landscape with craters and volcanoes; when there is, for example, a major shift in the priorities of a healthcare system such as with value-based care payments and pandemics, suddenly the software we create has to prioritize the more important clinically relevant findings based on the user's or clinical provider's needs. To preface this discussion and introduce a concept to which we will return later in this text, the end user of the software, whether a doctor, a nurse, or a patient, is often not the customer who purchases the product. This dichotomy between the "buyer" of medical software and the main "users" of a particular software make the design of the software even more challenging.

In addition to understanding the end users, we need to understand information inter-exchange standards. To put this in a different way, we need to answer the question: What "languages" do the other software systems with which we need to interact speak? This will be an important component of the design of our software,[1] as it defines some of the software's external interfaces. Our software, in addition to the direct user interface that allows our user to obtain, enter, and manipulate information within it, will almost certainly need to be able to obtain information from and store information in external systems. This information exchange will be a necessary component of our software requirements and design. We will briefly discuss two of the major standards, HL7 and DICOM, in Section 3.3. These two standards cover a large majority of the electronic health record and imaging software systems.

When operating in a healthcare environment, one also needs to be mindful of data privacy and security regulations. These issues must be considered early in any software design process. We also need to be mindful of cybersecurity concerns.

3.2 The Users: Patients, Doctors, and Everybody Else

3.2.1 The US Healthcare System

Unlike many other countries, the United States[2] has, for mostly historical reasons, a complex and often hard-to-understand system of providing healthcare.[3] In the United States, beginning in the 1930s, healthcare benefits were provided primarily via health insurance paid for by employers as part of worker compensation. In summary, there are three major components in the system: (1) the patients, (2) the healthcare delivery organizations (doctors, hospitals, etc.), and (3) the insurance companies that provide insurance policies.

Insurance is also provided by government agencies such as Medicare for the elderly and Medicaid for people with low incomes. In certain cases, individuals (for example, self-employed workers) may purchase insurance directly[4] from insurance companies. Finally, there is a separate system of healthcare for military veterans, run by the Veterans Administration (VA). This agency operates separate "VA" hospitals for this purpose. The VA system is the closest the United States has to a single-payer system of healthcare.[5]

The Stakeholders
Points of Care These usually refer to the locations (physical or virtual, such as in the case of telehealth) where a person obtains medical care. Please note the vast array of setups that are available. Clearly, hospitals and physicians' offices are the major players, but a significant amount of care is provided in places such as rehabilitation facilities, assisted living facilities, and long-term care facilities. These have a very different relationship with the end user, or patient and/or patient family member. The "customer" (patient) may actually be a long-term resident in the facility as opposed to a short-term visitor.

When the care is paid for by insurance, the customer and end user differ, as discussed above, but when the procedure is an "out-of-pocket cost," such as dental care or a cosmetic procedure, the end user and the customer are the same.

The reason that the point of care has become such a hotbed of interest for medical software development is that more and more patients are demanding "just in time" care. Just as we are used to ordering something online and seeing it arrive (hopefully) within a day or two, patients are no longer willing to wait a month for a lab test or several months for an appointment. Of course, this discussion uses the US healthcare system as an example; certain other countries have set the expectation that point of care does not translate into immediate attention.

Payers These are entities that pay for the care. In some countries, such as Canada, there is a single-payer system where all payments are performed by the government. In the United States, the system is much more complex (as described in the previous section), and there are a variety of primarily medical insurance organizations that perform the bulk of these payments. In addition, for some procedures, the consumer/patient may pay directly for services. While this is a vast oversimplification, let it suffice to say that in the United States, if a service is covered by a patient's medical insurance, it is much more likely to be accessed and utilized. This becomes an important consideration when designing medical software whose use is not covered by insurance. Such software may impact the provider or patient considerably, but does not impact the bottom line.

Clinical Support In addition to primary caregivers, there is a vast array of clinical support organizations that perform specialized functions. Probably the most common such entity is the pharmacy, where patients go to obtain medications. Other such facilities include clinical labs, where one goes to have blood-work done, and standalone imaging facilities. Some of these facilities may also exist inside large hospitals. Within the larger health systems, clinical support certainly includes the large electronic health record and much of the medical software, and the experts on the software have become an increasingly valuable part of the healthcare team.

Business Associates These provide additional support to a point of care. One good example is billing processors, who handle the billing aspect for smaller providers (e.g. dentists' offices). These are specialized operations and enable smaller practices to outsource some of their operations.

Other Entities These include insurance companies, law firms,[6] auditors, pharmaceutical/medical device companies, and contract research organizations (CROs – these work with healthcare providers and patients to manage clinical trials for new drugs/procedures). Many of these entities, such as CROs, are multi-billion-dollar operations that have their own complex medical software. Such software often has to work with the hospital's electronic health record, so interoperability is paramount as well.

IT Services The final category is information technology services. While many large hospitals may have in-house operations, increasingly organizations rely on external providers for some of these services. Data storage and backup, for example, may rely on cloud providers. In addition, such companies provide patients with access to information via web portals and mobile applications.

Note: Software developers may find this list of stakeholders useful as they try to understand a new situation, such as when identifying user needs – see Chapter 11. The relationship that, for example, a provider has with a patient (an occasional visitor as opposed to a resident) will impact what the needs of the provider are. The key point to remember is that the users are more than either the doctors or the patients. Our user may well be a legal company that tries to assess whether a physician provided appropriate care.

3.2.2   Healthcare Systems in the Rest of the World

Globally, healthcare systems are often complicated in their system of payment. Many nations, such as the United Kingdom, have a single-payer system with the government serving as the centralized payer of healthcare services. In other nations, such as India and Singapore, there also exist government hospitals, with the government paying for basic healthcare services. These, however, are often two-tier systems with a burgeoning private sector in which the speed, and some may argue the quality, of care is better and available for a price. When developing medical software and selling the product within any geographic region, it is important, as pointed out later in this text, to understand the economic system of payment as well as the organizational entities responsible for implementation. There are few countries with only private healthcare systems, although some of the poorer nations certainly have very basic healthcare only available for its citizens free of charge.

A national or centralized approach can be beneficial as purchasers of novel software programs and analytics as they have a coordinated electronic platform of all the patients, records. Conversely, government systems such as in the Scandinavian countries, for example, are often slow to adopt technologies and are very cognizant of the costs and benefits of the software programs, as well as privacy concerns around the security of the medical software design. In fact, most developed countries have a national health plan that covers its citizens with basic healthcare. In the UK, in particular, the National Health Service is even more intimately involved in the delivery of healthcare, making recommendations on digital health applications, hosting innovation forums, and assessing the costs of additional services quite carefully.

In contrast, the majority of healthcare systems globally function with a two-tier system of services, with the private hospitals and clinics having a myriad of electronic health records and medical devices that are often not inter-connected. Unfortunately, in addition to disparate patient data, this makes for a very complex myriad of healthcare systems in which the data is often less centralized and the roll-out of massive software integration much more challenging. Examples of this include the United States, European nations such as Germany, and many Asian and African nations such as Singapore, India, and Egypt. In recent years, China has emerged as an interesting case study in healthcare systems as the centralized form of government and tight control over supply of doctors and clinical providers has led to an increase not just in healthcare services, but also training programs, with opening of medical and nursing schools. This text will cover some of the complex regulatory and systems principles in these regions, but certainly these topics are much more complex than can be covered in this introductory text.

3.2.3   Who Is My Customer?

A medical software engineer needs to have an appreciation of the healthcare setup, as this will affect the design and development of the software. A key aspect of this process is identifying what the users' needs are; consequently, one needs to identify who the users effectively are before even beginning the process of identifying user needs and potential solutions.[7] Some products will appeal to insurance companies (if they improve patient outcomes and thus reduce costs in the long run; hospitals also have outcome improvement expectations). Others may affect the efficiency of care and might therefore find support in hospitals that can use devices/software to improve equipment utilization, resulting in the ability to serve greater numbers of patients. Finally, some may appeal to patients directly, resulting in their choosing (if they can) to go to a particular health facility for care if it has a device that they believe will provide, for example, superior outcomes or superior patient comfort.

3.3   Clinical Information Technology

The term clinical information technology (IT) refers to the backbone computer and network infrastructure (including databases) that provide computing services to support the operation of a clinical entity. What are the major systems a software designer might need to interact with in a clinical IT environment? The first is probably the electronic health record (EHR) system – this is also commonly referred to as the electronic medical record (EMR) system. The transition from paper records (large folders containing multiple pieces of paper) to the EHR has been one of the central challenges in the application of IT to medicine. We will discuss this further in Section 3.3.1. The second major database system is the picture archive and communication system, or PACS [208]. The PACS is the medical image database system and is probably the largest shared database by size in most hospitals.[8] We will discuss the PACS and interfacing with it in Section 3.3.2.

    A major component of clinical IT operations is ensuring data privacy and security. This includes HIPAA and GDPR compliance (depending on the regulatory zone). A major part of such compliance is cybersecurity – ensuring that our data and computers are protected from unauthorized access (primarily but not exclusively over the Internet). We discuss these issues in Section 3.4.

3.3.1   Electronic Health Records

A good definition of EHR is provided by the Healthcare Information and Management Systems Society (HIMSS) (as quoted by Atherton [14]):

The Electronic Health Record (EHR) is a longitudinal electronic record of patient health information generated by one or more encounters in any care delivery setting. Included in this information are patient demographics, progress notes, problems, medications, vital signs, past medical history, immunizations, laboratory data and radiology reports. The EHR automates and streamlines the clinician's workflow. The EHR has the ability to generate a complete record of a clinical patient encounteras well as supporting other care-related activities directly or indirectly via interfaceincluding evidence-based decision support, quality management, and outcomes reporting.

It is only a mild understatement that the transition from paper records to EHRs met with a negative reaction from physicians and other providers. Doctors, for the most part, hated and still hate the transition from paper to electronic records. In an article that appeared in *The New Yorker* [96], Dr. Atul Gawande, a surgeon and public health researcher, describes how his initial optimism about the benefits of moving to a central EHR gradually evaporated. To quote a particular poignant section:

But three years later I've come to feel that a system that promised to increase my mastery over my work has, instead, increased my work's mastery over me. I'm not the only one. A 2016 study found that physicians spent about two hours doing computer work for every hour spent face to face with a patientwhatever the brand of medical software. In the examination room, physicians devoted half of their patient time facing the screen to do electronic tasks. And these tasks were spilling over after hours. The University of Wisconsin found that the average workday for its family physicians had grown to eleven and a half hours. The result has been epidemic levels of burnout among clinicians. Forty per cent screen positive for depression, and seven per cent report suicidal thinking almost double the rate of the general working population.

The proponents of EHRs argue, however, that in digitizing patient information in a central system EHRs can allow this information to be used to assess current practices and develop more evidence-based approaches to medical treatment. The last statement in the HIMSS definition alludes to this when it states that EHRs should also provide functionality for "supporting other care-related activities directly or indirectly via interfaceincluding evidence-based decision support, quality management, and outcomes reporting." This was a key part of the rationale provided by the original legislation for this move. For example, if all patient information is stored digitally, one can more easily ask questions such as "What were the outcomes for patients in a certain age range that were given a particular medication for a particular condition?" This type of query was almost impossible to pursue in the era of paper records or even decentralized electronic records.

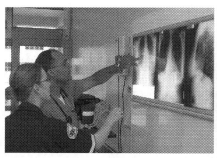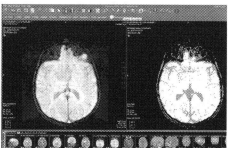

Figure 3.2 Evolution of clinical radiology viewing stations. *Left:* A standard lightbox for viewing clinical X-ray images. This was the most common way to view X-rays until the adoption of digital imaging, beginning in the late 1990s (image from the United States Navy, public domain). *Right:* A modern PACS viewer (Visage). This is the system currently in use at Yale.

Interfacing with the EHR is a nontrivial affair [258]. The most common standard (or family of standards) is HL7 [110], which is implemented by many EHR providers. One definition of HL7 is that it is "a protocol designed for signaling changes to a patient's record using short messages" [237, 258]. Many EHR providers document how to interface with their systems using HL7 standards. For example, Epic [59], one of the biggest providers of EHR systems in the United States, has a web page [60] that describes how to use HL7 (version 2) to communicate with Epic's EHR system. To be more specific: An external software system can use Epic's "Incoming Clinical Observations" interface to add observations to the database. The page (as of 2020) lists 80 such interfaces, including patient administration, appointment scheduling, billing, pharmacy orders, and imaging orders.[9]

3.3.2 Imaging Databases: PACS and DICOM

Imaging is a major source of information in medicine. Originally, imaging technologies were analog (e.g. film-based X-rays) and were viewed using lightboxes (see Figure 3.2, left). In addition, even digital images (e.g. X-ray computed tomography images, or CT) were printed on film for viewing using these devices.[10]

With the advent of tomographic imaging there was a progressive move toward digital viewing of images (see Figure 3.2, right) on computer workstations. The reason for this is that tomographic images are by construction digital, as they are produced using software-based reconstruction techniques (e.g. MRI, X-ray CT, PET, SPECT). This transition went hand-in-hand with the development of the PACS, beginning in the early 1980s [114]. The PACS is the main digital imaging database in a hospital and stores all the digital medical images acquired as part of patient care. Typical PACS in large hospitals store petabytes of data.

The development of the modern PACS has revolutionized the practice of clinical radiology. Radiologists can now read images remotely over the Internet. Many small clinical facilities, in fact, outsource their radiology practice to other locations (even other countries) via teleradiology services.

A term that is used often as a synonym for PACS is DICOM. This stands for Digital Imaging and Communications in Medicine [208], and it is the communication standard used to transmit and receive information between the PACS and other devices such as imaging scanners, radiology workstations, and EHR systems. The DICOM standard [171] consists of multiple parts, each covering different aspects of the communication protocol. It is updated several times a year. In addition to the standard, there are some well-supported open-source implementations such as DCMTK [196]. A recent extension of DICOM is DICOMWeb, an implementation of DICOM that uses (according to the standard) "RESTful services, enabling web developers to unlock the power of healthcare images using industry-standard toolsets" [172]. This may be easier to implement than standard DICOM.

Any software that displays or analyzes medical images will need to have integration capabilities before it can be successfully integrated into a clinical setting. The standardization of the imaging tools has allowed healthcare providers to share images much more easily than when using physical discs or printed X-rays.[11]

A note on medical images: Unlike standard camera images, medical images contain a significant amount of metadata that changes the way they are displayed and interpreted. For example, pixels in medical images need not be square (i.e. the image may have different x- and y-resolutions). A key aspect of the metadata (contained in the DICOM header) is the image orientation that determines the position of each pixel in 3D space. This allows one to, for example, determine whether the x-axis is pointing from the left to the right or vice versa, or whether the image was acquired using a nonstandard orientation. Failure to account for image orientation could result in images being wrongly displayed, with potentially catastrophic implications if, for example, the image is being used for surgical planning.

3.4 Data Privacy

Inevitably, medical software handles, creates, and processes information about patients. This information is now increasingly subject to data privacy and security legislation, which imposes significant constraints on how software is designed.

It is safe to say that the words HIPAA [49] and GDPR [65] are major sources of fear and angst for medical software developers. These are sets of laws and regulations to prevent unauthorized use of or access to personal data. Fundamentally, these regulations create an environment in which the software design must incorporate data protection by default – this is stated explicitly in Article 25 of the GDPR. Failure to do so can have serious financial implications/penalties, as described in a report by ANSI [9]. This is particularly the case with the GDPR, which applies to all types of entities (not just specific entities handling medical information), regardless of where they are located.

The rest of this section consists of a discussion of HIPAA, data anonymization, and the GDPR, and concludes with a discussion of cloud security issues.

3.4.1 HIPAA

As described on the website of the US Department of Health and Human Services, one of the provisions of the *Health Insurance Portability and Accountability Act (HIPAA, 1996)* [49] is "to adopt national standards for electronic health care transactions and code sets, unique health identifiers, and security. At the same time, Congress recognized that advances in electronic technology could erode the privacy of health information. Consequently, Congress incorporated into HIPAA provisions that mandated the adoption of Federal privacy protections for individually identifiable health information."

One often hears two related terms in relation to HIPAA. These are **PHI** or private health information and **ePHI** or electronic private health information. Part of the definition of PHI is that it is *individually identifiable* private information.[12] As stated in the HHS webpage:

Protected health information is information, including demographic information, which relates to:

the individual's past, present, or future physical or mental health or condition,

the provision of health care to the individual,

or the past, present, or future payment for the provision of health care to the individual, and that identifies the individual or for which there is a reasonable basis to believe can be used to identify the individual. Protected health information includes many common identifiers (e.g. name, address, birth date, Social Security Number) when they can be associated with the health information listed above.

This is a very broad definition. For most software applications, one must consider the usual list of 18 identifiers that constitute ePHI:[13]

- name;
- address (all geographic subdivisions smaller than state, including street address, city, county, ZIP code);
- all elements (except years) of dates related to an individual (including birth date, admission date, discharge date, date of death, and exact age if over 89);
- telephone numbers;
- fax number;
- email address;
- Social Security number;
- medical record number;
- health plan beneficiary number;
- account number;
- certificate/license number;

- any vehicle or other device serial number;
- device identifiers or serial numbers;
- web URL;
- Internet Protocol (IP) address numbers;
- finger or voice prints;
- photographic images;
- any other characteristic that could uniquely identify the individual.

Essentially this means that when storing any information about the patient that includes any of the above 18 items, one must be careful to ensure that this data is not accessed by unauthorized personnel. Typically this requires that the data is stored under "lock and key," – that is, on encrypted, password-protected computers. Unauthorized access to data, especially when the number of patients involved is large, is a serious legal issue that can lead to substantial costs [9].

From a practical standpoint, a medical software designer has to imagine that if one can have access to any 2–3 pieces of information, such as name, email address, and photographic images, one can make considerable inferences about the patient's health, as well as any medical procedures they may have had. Hence, such pieces of information should be kept as secure and as separate as possible.

Originally only covered entities [9] were subject to HIPAA. Since then, additional regulations have been issued that extend these protections. *The Genetic Information Non-Discrimination Act of 2008* added special privacy protections to genetic information.[14] The *American Recovery and Reinvestment Act of 2009*,[15] which also provided funding to encourage the adoption of EHRs (see Section 3.3.1), added additional protections for individual privacy rights, and penalties for those violating these rights. It also extended HIPAA privacy and security standards (and penalties) to "business associates" of covered entities. Examples of such business associates include companies involved in claims processing, medical transcription companies, and companies involved in data conversion and de-identification.

From a software designer's perspective, this leads to an increased incentive not to directly store anything other than the bare minimum of patient data and to store that minimum in such a way that it does not constitute ePHI. One common technique is to identify data (e.g. an image) using a private identifier such as an arbitrary number that is unrelated to the patient. For example, instead of using the patient's name, we can substitute "Patient 002." If needed, the mapping of this private identifier ("Patient 002") to the actual patient is handled by a separate, more secure computer to which access is highly restricted.

3.4.2   Anonymization

The US HHS has a useful guidance document on de-identification, titled *Guidance Regarding Methods for De-identification of Protected Health Information in Accordance with the Health Insurance Portability and Accountability Act (HIPAA) Privacy*

Rule [48]. This describes the two methods for meeting the HIPAA de-identification standard. These are "expert determination" and "safe harbor." In the case of expert determination, a domain expert examines the data and applies statistical or scientific principles to demonstrate that the risk of someone receiving the data being able to identify an individual is very small. In the case of safe harbor, the 18 types of identifiers are removed from the data and, as such, an individual can not be re-identified. To give an example, in using the second approach, the age of a subject is completely removed, whereas in the first approach we could specify an age range (e.g. 35–44) instead of the actual age.[16] If there are many subjects in the data set that belong to this age range, then statistically it will be hard to identify a particular individual based on this information.

3.4.3 GDPR

GDPR stands for *General Data Protection Regulation* [65] and is a regulation issued by the EU in 2016 and adopted in 2018. It represents a much broader set of rules than HIPAA. Unlike HIPAA, which only applies to specific entities (e.g. hospitals) that handle protected health information in the United States, GDPR addresses additional sensitive personal data (e.g. religion, ethnic origin) and applies to all entities holding or using personal data of EU residents, regardless of the location of the entities. When GDPR went into effect, almost every website had a pop-up notice informing users of what information the website/organization had and asking for consent. Many US-based entities simply blocked access to users coming from the EU to avoid being entangled in this type of issue.

As defined in Article 4 of the GDPR regulations:

"personal data" means any information relating to an identified or identifiable natural person ("data subject"); an identifiable natural person is one who can be identified, directly or indirectly, in particular by reference to an identifier such as a name, an identification number, location data, an online identifier or to one or more factors specific to the physical, physiological, genetic, mental, economic, cultural or social identity of that natural person.

As one can easily see, this is a broader list than what constitutes PHI in HIPAA. The goal of GDPR is essentially to force entities that hold personal data to design privacy protections as a core feature of their operations, and not simply as an added security layer. This is made explicit in Article 25 of the GDPR, "Data protection by design and by default." Before quoting from this article, we first note that the GDPR defines the term controller as "the natural or legal person, public authority, agency or other body which, alone or jointly with others, determines the purposes and means of the processing of personal data." With this definition in mind, the following quotes from this article illustrate what the requirements are for software designers:

the controller shall, both at the time of the determination of the means for processing and at the time of the processing itself, implement appropriate technical and organisational measures, such as pseudonymisation, which are designed to implement data-protection principles, such as data minimisation, in an effective manner and to integrate the necessary safeguards into the processing in order to meet the requirements of this Regulation and protect the rights of data subjects.

The controller shall implement appropriate technical and organisational measures for ensuring that, by default, only personal data which are necessary for each specific purpose of the processing are processed. That obligation applies to the amount of personal data collected, the extent of their processing, the period of their storage and their accessibility. In particular, such measures shall ensure that by default personal data are not made accessible without the individual's intervention to an indefinite number of natural persons.

The take-home lesson is that the way one stores private information should be designed in from the beginning and not added as an afterthought.

3.4.4   Data Security and Privacy in the Cloud

Historically, most data storage in a clinical environment was housed in secure data centers, protected by firewalls that prevented unauthorized access. This was, and for the most part still is, certainly the case in large hospitals, where all remote access happens via encrypted virtual private network connections. Increasingly, however, many application providers rely on cloud services such as Amazon Web Services or Microsoft Azure (see Section 9.4) to distribute software solutions for healthcare providers. In this setup, the data physically sits on a server that may be located far away from the hospital, potentially even in another country.

An example of a cloud-based service is a cardiac image analysis package by Arterys Inc. [11] that was cleared by the FDA in 2017 [10]. When using this software, the physician uploads the data to an external provider's computer system. The data is analyzed by this system, which then sends a report back to the physician. Clearly this "data" transmission out of the protected setting raises data security and privacy issues per HIPAA and GDPR. The solution that Arterys uses is to de-identify the data on upload and then re-identify it when it is downloaded back to the original source. This avoids the transmission of PHI, reducing the risk of data "leakage."

3.5   Cybersecurity

Cybersecurity is a complex topic in its own right and many high-profile cases have recently made headlines – see Figure 3.3 for two examples. Our goal here is to provide a brief introduction to this topic, particularly as it relates to medical software design, and to provide pointers to further reading. In an increasingly network-connected world, cybersecurity is a critical component of ensuring data privacy and the safety of our software. Given the increasing use of AI/ML modules and their susceptibility to adversarial attacks (see Section 5.2.4), cybersecurity may become even more central in the deployment of medical software.

The IMDRF defines cybersecurity as "a state where information and systems are protected from unauthorized activities, such as access, use, disclosure, disruption,

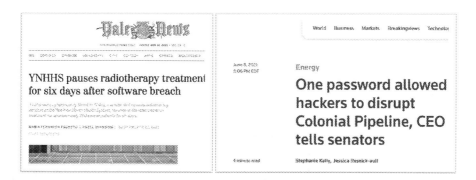

Figure 3.3 Two high-profile cybersecurity incidents This figure shows news coverage of two high-profile cybersecurity incidents. On the left, the story describes the case in which issues with a cloud-based system run by a vendor called Electra resulted in a pause in radiotherapy treatment for patients at the Yale New Haven Health System (YNHSS) for six days. Image of the Yale Daily News, April 20, 2021. The story on the right relates to the disruption of a major oil pipeline in the United States as a result of a hacking incident. Image of Reuters, June 8, 2021.

modification, or destruction to a degree that the related risks to confidentiality, integrity, and availability are maintained at an acceptable level throughout the life cycle. (ISO-81001-1)" [129].

3.5.1 IMDRF Guidance

For medical software, the IMDRF [129] lists a number of design principles for medical device manufacturers. These should be followed in addressing cybersecurity needs. We summarize some of this material below:

- **Secure communications** over interfaces such as Ethernet (including wireless), Bluetooth, and USB. In particular, all such communications should be secured. Communications with less secure (e.g. home networks and legacy devices) should be validated. Consideration should be given to the use of authentication and encryption, as needed.
- **Data protection**, which may involve consideration of encryption techniques and, in particular, require that hard drives be encrypted at the system level.
- **Device integrity**, which may involve ensuring that the system is not changed in an unauthorized manner, or by malicious code (e.g. viruses, malware).
- **User authentication**, including the use of appropriate access controls and secure passwords.
- **Software maintenance**, which should include procedures for regular updates of software and updates to the operating system and other third-party tools.
- **Physical access restrictions** can be some of the most effective means of defense. This may involve physical locks (or locked rooms).
- **Reliability and availability** considerations include features that will allow the device to resist and survive cybersecurity attacks.

The document also has a recommendation that security requirements be identified during the requirements phase of a project, and be included in the appropriate documents – that is, the system requirements document (see Chapter 12). In addition, one must provide appropriate end user security documentation, perform risk management for cybersecurity, and have a plan for detecting and responding to threats [234].

3.5.2   FDA Guidance on Cybersecurity

The FDA's recent draft cybersecurity guidance [87] divides medical software into two tiers:

1. **Tier 1** – higher cybersecurity risk;
2. **Tier 2** – standard cybersecurity risk.

A device is classified as Tier 1[17] if it can connect to another medical or non-medical product, or to a network, or to the Internet *and* when a cybersecurity incident affecting the device could directly result in patient harm to multiple patients – this is clearly the case of the Electra device shown in Figure 3.3. Otherwise the software/device is classified as Tier 2.

The guidance notes that examples of Tier 1 software "include but are not limited to, implantable cardioverter defibrillators (ICDs), pacemakers, left ventricular assist devices (LVADs), brain stimulators and neurostimulators, dialysis devices, infusion and insulin pumps, and the supporting connected systems that interact with these devices such as home monitors and those with command and control functionality such as programmers."

In design guidance, a "trustworthy device" has the following attributes:

1. The device is reasonably secure from cybersecurity intrusion and misuse.
2. The device has a reasonable level of availability, reliability, and correct operation.
3. The device is suited to performing its intended functions.
4. It adheres to accepted security procedures.
5. It was designed in a way that was intended to be consistent with the NIST (National Institute of Standards and Technology) framework [187].

The last item in this list points to the cybersecurity framework produced by the NIST [187]. This is a great introduction to this topic and divides cybersecurity into five areas:

1. **Identify:** Understand your environment and what the risks are to systems, data, and people.
2. **Protect:** Implement safeguards (ahead of time) to limit the impact of a cybersecurity event.

3. **Detect:** Have procedures to identify incidents. This involves having some form of continuous monitoring system in place.
4. **Respond:** Have the ability to contain the impact.
5. **Recover:** Have plans to restore any services/capabilities that were affected by an incident.

The FDA Guidance amplifies these areas. For example, under "Identify and Protect," we have three main areas: (1) prevent unauthorized use; (2) ensure trusted content by maintaining code, data, and execution integrity; and (3) maintain confidentiality of data. A good summary of the points made there might be "authenticate and encrypt." One must ensure that access is only granted to appropriately authorized users or software (at different levels of permissions) and that sensitive information is encrypted both *when stored* and *when transmitted.*

We note here that both of the interfacing standards we discussed in Section 3.3, HL7 and DICOM, for the most part require no authentication and do not use encryption. In most hospitals, the solution is to place all such servers behind a firewall to provide a secure space (where access is limited by authentication and use of virtual private networking software). One problem with this arrangement is that the authentication mechanism is not granular (as repeatedly emphasized in the guidelines), and once authenticated a user has complete access to the secure network, as opposed to the more ideal case of having access to only the minimal amount of resources required by that individual to perform their duties.

3.5.3 Concluding Thoughts

Cybersecurity Involves All Stakeholders Finally, we need to emphasize the importance of having all involved stakeholders participate in ensuring cybersecurity. As the regulations from Singapore [234] state:

Cybersecurity . . . cannot be achieved by a single stakeholder, it requires the concerted effort of diverse stakeholders (government agencies, manufacturers, healthcare institutions, users of medical devices). Continuous monitoring, assessing, mitigating and communicating cybersecurity risks and attacks requires active participation by all stakeholders in the ecosystem.

Many high-profile cybersecurity incidents are caused by user error or weak passwords or users falling victim to phishing attacks (e.g. official-looking emails asking users to reset their passwords, such as case of the hacking of the Hilary Clinton presidential campaign in 2016 [28]), and the solutions may need to come primarily from increased user education and training.

Cybersecurity and Risk Management Cybersecurity is an integral part of the risk management process for medical software. We will discuss risk management in more detail in Chapter 5. The same basic process for risk management directly applies to the evaluation and mitigation of cybersecurity threats as well. Similar to other risk management processes, cybersecurity concerns should be addressed at each stage of the software life cycle.

3.6   Conclusions

In designing medical software one needs to account for a number of additional constraints over and above those resulting from the regulatory agencies that perform the review (such as the FDA). The first is the ability of the software to function within the healthcare environment. This requires an understanding of the stakeholders and users, and also the ability to interoperate with existing large-scale IT infrastructure, such as EHRs and PACS. In addition, one must account for the increasing worldwide legislative emphasis on data privacy and security that is a direct response to the challenges of our increasingly interconnected world. The same interconnectivity also presents cybersecurity challenges that the software developer must be aware of. This includes ensuring the use of authentication and encryption mechanisms to limit access to sensitive systems and data.

3.7   Summary

- Medical software operates in a complex healthcare environment with a myriad of stakeholders and types of users. An important issue is the distinction (and often conflict) between users (those that use the software, e.g. doctors) and buyers (those that will pay for the software, e.g. a hospital system), which creates additional challenges in meeting their separate expectations.
- The software may need to connect and interoperate with large-scale clinical software infrastructure such as EHRs and PACS. This may require the developer to implement functionality for HL7 and DICOM information inter-exchange.
- Data privacy and security must be accounted for at the beginning of the software process and not as an afterthought. The same applies to issues related to cybersecurity. Data anonymization techniques can be very helpful in alleviating some of these problems.

RECOMMENDED READING AND RESOURCES

J. Atherton provides a concise history of the development of the modern EHR.

J. Atherton. Development of the electronic health record. *Virt Mentor*, 13(3):186–189, 2011.

There was a rocky transition from paper to electronic health records, which elicited much negative reaction from physicians. The story is told well in the following article:

A. Gawande. Why doctors hate their computers. *The New Yorker*, November 5, 2018.

A good description of the issues with Electronic Health Records in general and HL7 in particular can be found in the following book:

F. Trotter and D. Uhlman. *Hacking Healthcare: A Guide to Standards, Workflows, and Meaningful Use.* O'Reilly Media, 2011.

For actual examples on using HL7, see the following white paper:

R. Spronk. Ringholm whitepaper: HL7 message examples: version 2 and version 3, November 16, 2007. www.ringholm.de/docs/04300_en.htm.

An excellent description of all things PACS and DICOM can be found in:

O.S. Pianykh. *Digital Imaging and Communications in Medicine (DICOM): A Practical Introduction and Survival Guide.* Springer, 2nd edn., 2012.

A recent report by the American National Standards Institute (ANSI) that can be freely downloaded from their website details the potential legal costs of unauthorized patient data access:

American National Standards Institute (ANSI) and the Santa Fe Group/Internet Security Alliance. The financial impact of breached protected health Information: A business case for enhanced PHI security. Technical report, ANSI, 2012.

One of the standard introductions to cybersecurity is the framework produced by the US NIST:

National Institute of Standards and Technology. Framework for improving critical infrastructure cybersecurity, April 16, 2018. www.nist.gov/cyberframework.

There are a number of guidance documents on cybersecurity from various regulators/countries. Some of these are listed below

FDA. "Cybersecurity guidelines." www.fda.gov/medical-devices/digital-health-center-excellence/cybersecurity#guidance.

EU Medical Device Coordination Group. MDCG 2019-16: guidance on cybersecurity for medical devices, July 2020.

Republic of Korea Ministry of Food and Drug Safety, Medical Device Evaluation Department. Guideline on review and approval for cybersecurity of medical devices, November 4, 2020.

An recent excellent and detailed coverage of cybersecurity and medical devices can be found in:

A. Wirth, C. Gates, and J. Smith. *Medical Device Cybersecurity for Engineers and Manufacturers.* Artech House, 2020.

Microsoft's threat modeling page (part of their security documentation/tools) is also an excellent and freely available online resource. It includes the definition of the STRIDE model, which categorizes threats as **s**poofing, **t**ampering, **r**epudiation, **i**nformation disclosure, **d**enial of service, and **e**levation of privilege. This is a great organizing structure for discussing security issues. Microsoft also makes available a free threat modeling tool from the same website.

NOTES

1. This is a critical component of the intended use environment. We will discuss this concept in Chapter 12 as part of creating our system requirements specification.

2. This section is useful also for those readers based outside the United States, as the United States is one of the largest markets for medical devices/software and an understanding of the perplexing US healthcare setup is important for developing products for the US market.

3. One may also point out that the exact structure of healthcare delivery in the United States is a matter of ongoing political controversy. We will not concern ourselves here with the question of "what the system ought to be like" but simply focus on "what the system is like in 2021" and how it affects medical software.

4. The long saga of the creation of the federal online marketplace for directly purchasing policies from insurance companies is described in the vignette presented in Chapter 19.

5. In a single-payer healthcare system, such as that in Canada, the government is responsible for covering all medical expenses.

6. There is a significant medical malpractice component in many law firms, where patients can sue providers for damages due to substandard care.

7. The dichotomy between users and buyers is important here. The software designer may meet with what she perceives as the users (i.e. the doctors). However, these users are not actually the ones that will pay for the software (i.e. they are not the buyers). The buyers are the employers of the users, such as the hospitals/healthcare systems. These buyers have budgetary constraints and are more likely to be willing to pay for something if it can demonstrate definite improvements in patient health outcomes.

8. Modern medical images are significantly larger than text and audio recordings. For example, a single brain anatomical MRI can take 10–20 MB of storage (uncompressed).

9. On a related topic, it has recently become clear that hospitals and health systems own a vast array of population health data within their EHR, and controlling the flow of that data may be important for the medical software designer to be cognizant of as well.

10. Some clinical images (e.g. clinical pathology slides obtained from biopsy samples) are still commonly viewed in an "analog" manner using microscopes. Analog technology still has some significant advantages in terms of storage and resolution.

11. The development of the PACS has allowed providers across geographic regions to share images of tumors or confer on difficult cases; however, imaging integration and viewing within EHRs is still fraught with the same pitfalls as the general interoperability issues within EHRs.

12. A related, more general, term is personally identifiable information (PII). See the NIST report titled *Guide to Protecting the Confidentiality of Personally Identifiable Information (PII)* [167] for more details.

13. This list was obtained from the website of the Yale HIPAA office.

14. One can imagine, for example, a person having difficulty obtaining life insurance if the insurer were able to determine that she had a genetic issue that predisposed her to serious breast cancer.

15. This was colloquially known as the "Stimulus" Act and was passed in the aftermath of the 2007 financial crisis. It included significant health provisions, including funding for EHRs. We discuss these in more detail in Section 3.3.1.

16. Keeping the information in the data may allow more sophisticated statistical analyses (e.g. grouping subject data by age range) as opposed to removing it completely.

17. This classification is independent of the general device risk classification (e.g. Class I, II, or III).

4   Quality Management Systems

INTRODUCTION

This chapter provides an overview of quality management systems (QMS). There is an increasing emphasis of regulators on the "organization" as opposed to the "product," which places an even greater emphasis on the use of a QMS. We first introduce what a QMS is (Section 4.1) and provide some regulatory background, including a discussion of the recent FDA precertification program. Next, we discuss various international standards that apply (Section 4.2). The core of this chapter is an extended discussion of the IMDRF QMS regulatory document (Section 4.3), which provides an excellent and SaMD-specific description of this important topic. The chapter concludes with a brief discussion (Section 4.4) of the implications of quality systems (or rather their absence) on research and scientific software and how failure to have/adhere to a QMS (Section 4.5) can lead to serious problems.

4.1   Introduction: The Organization Matters

The goal of this chapter is to introduce the concept of quality management systems (QMS). Our hope is to provide the junior developer starting out in a medical device/medical software company with enough information to understand the constraints under which the managers of the company are operating as they manage both individual projects and the company itself. This context will hopefully help this developer to better understand why certain things are done the way they are and where all the paperwork comes from!

What is a QMS? The American Society for Quality [13] defines a QMS as:

a formalized system that documents processes, procedures, and responsibilities for achieving quality policies and objectives. A QMS helps coordinate and direct an organization's activities to meet customer and regulatory requirements and improve its effectiveness and efficiency on a continuous basis. ISO 9001:2015, the international standard specifying requirements for quality management systems, is the most prominent approach to quality management systems. While some use the term "QMS" to describe the ISO 9001 standard or the group of documents detailing the QMS, it actually refers to the entirety of the system. The documents only serve to describe the system.

4.1.1   Regulatory Background

An organization's QMS defines the organizational culture of the company. It captures in a set of documents all the policies required for the organization to function. Having a QMS is practically mandatory for FDA clearance. The FDA Quality System Regulations [69][1] (QSR) explicitly state that one must have a quality system – see especially Subpart B of that document.

A detailed regulatory statement of what a QMS should include can also be found in the EU regulation EU 2017/745 on medical devices [64]. Article 10 of this regulation provides the following list:

(a) a strategy for regulatory compliance, including compliance with conformity assessment procedures and procedures for management of modifications to the devices covered by the system;

(b) identification of applicable general safety and performance requirements and exploration of options to address those requirements;

(c) responsibility of the management;

(d) resource management, including selection and control of suppliers and sub-contractors;

(e) risk management as set out in Section 3 of Annex I;

(f) clinical evaluation in accordance with Article 61 and Annex XIV, including PMCF ("Post-market Clinical Follow-up");

(g) product realisation, including planning, design, development, production and service provision;[2]

(h) verification of the UDI ("Unique Device Identifier") assignments made in accordance with Article 27(3) to all relevant devices and ensuring consistency and validity of information provided in accordance with Article 29;

(i) setting-up, implementation and maintenance of a post-market surveillance system, in accordance with Article 83;

(j) handling communication with competent authorities, notified bodies, other economic operators, customers and/or other stakeholders;

(k) processes for reporting of serious incidents and field safety corrective actions in the context of vigilance;

(l) management of corrective and preventive actions and verification of their effectiveness;

(m) processes for monitoring and measurement of output, data analysis and product improvement.

As shown above, a QMS covers all aspects of running a company, from resource and risk management to clinical evaluation and post-market follow-up. Typically, the components of a QMS are encapsulated in procedures that are described in more detail in standard operating procedure (SOP) documents. These are specific to each organization (though they often derive from a standard template). For example, a medical software company will probably have a "Managing Source Code" SOP.

Finally, we have the guidance of industry standards such as ISO 9001 [131] and ISO 13485 [132]. In addition, for medical software, the IMDRF *Document on Quality Management Systems* (IMDRF QMS) [127] is also exceedingly useful. In this chapter we will primarily follow the descriptions in the IMDRF document.

4.1.2   The FDA Precertification Program

In July 2019, the FDA introduced a pilot digital health software precertification program [85]. This *pilot* program is restricted to software as a medical device (SaMD) manufacturers [125]. Its goals are to:

help inform the development of a future regulatory model that will provide more streamlined and efficient regulatory oversight of software-based medical devices developed by manufacturers who have demonstrated a robust culture of quality and organizational excellence, and who are committed to monitoring real-world performance of their products once they reach the U.S. market.

This is reminiscent of the old adage: "Nobody ever got fired for buying IBM" (computers). Essentially, if the company establishes a strong reputation for quality, the FDA may begin (this is a pilot program) to trust the company and allow it to use a more streamlined process. The next paragraph reinforces this shift in emphasis from reviewing the product to reviewing the organization:[3]

This proposed approach aims to look first at the software developer or digital health technology developer, rather than primarily at the product, which is what we currently do for traditional medical devices.

4.2   Standards: ISO 9001, 90003, and 13485

4.2.1   Introduction

ISO 9001 [131] (last revised in 2015) is the international standard for QMS. It offers a framework for organizations to use to ensure quality in the design and manufacture of products and services. Companies can be certified as being ISO 9001 compliant. This type of certification sends a strong signal to both consumers and other customers that this particular organization follows best practices in its processes. ISO 9001 certification is not a one-time event. Companies continue to be audited by an ISO 9001 registrar (often once or twice a year) to ensure compliance.

In addition to this core document, ISO produces more specialized standards for particular industries. Of primary interest for our purposes in the medical device and medical software field are (1) ISO 13485 [132] and (2) ISO 90003 [134]. The first is a customization of ISO 9001 for medical devices, and the second is a customization for computer software. Medical software lies at the intersection of these two categories.

4.2.2   The Structure of ISO 9001 and 90003

ISO 9001 is a 40-page document. After an initial preamble and some definitions, there are seven major sections, as follows:

- Section 4 – Context of the Organization (1.5 pages);
- Section 5 – Leadership (1.5 pages);

- Section 6 – Planning (1.5 pages);
- Section 7 – Support (4 pages);
- Section 8 – Operation (7 pages);
- Section 9 – Performance Evaluation (3 pages);
- Section 10 – Improvement (1 page).

The section titles alone should give one a good idea of what a QMS should include. Cochran [34] notes in his review of the ISO 9001:2015 that there has been an increasing emphasis on organizational structure in the document as it has been revised over time, and a more or less explicit requirement that a company must have a strategic plan. So while ISO 9001 has a continuing focus on operation (i.e. how the products are being designed/implemented), it also has requirements for the whole organization. These include leadership, planning, providing support for the product teams, and, interestingly, performance evaluation and strategies for improvement. The understanding here is that all of these components contribute to the safety and efficacy of the product for the end user/consumer.

ISO 90003 follows the exact same structure as ISO 9001. It is a longer document (86 pages). Essentially, each section in ISO 90003 quotes and summarizes the corresponding section in ISO 9001 and adds commentary. For example, in Section 4.1 of ISO 9001 ("Understanding the Organization and Its Context"), we have the comment:

The organization shall determine external and internal issues that are relevant to its purpose and its strategic direction and that affect its ability to achieve the intended result(s) of its quality management system.

The organization shall monitor and review information about these external and internal issues.

ISO 90003 adds additional information, of which the following is a part:

Use of "Cloud" (i.e. network accessed systems provided by a third party) applications, tools and storage services. This can be of economic benefit as well as to provide for business continuity, but needs research to ensure there is no increased risk to the organization in using the cloud services provider.

In general, for software-specific projects, ISO 90003 provides useful additional commentary/information to the core ISO 9001 standard.

4.2.3 The Structure of ISO 13485

ISO 13485 [132] is a specialization of ISO 9001 for medical devices. It is also a relatively short document (46 pages). It is in the usual ISO format, a few introductory sections followed by the "major" sections, which are:

- Section 4 – Quality Management Systems – with an explicit focus on documentation requirements, quality manual, and controls for documents and records (3 pages);
- Section 5 – Management Responsibility – with subsections on management commitment, customer focus, quality policy, planning, responsibility, authority, and communication, and management review (2.5 pages);

- Section 6 – Resource Management – including human resources, infrastructure, and work environment (1 page);
- Section 7 – Product Realization (9 pages);
- Section 8 – Measurement Analysis and Improvement (4.5 pages).

ISO 13485 is based on ISO 9001 but expands to cover material more relevant to the medical world (e.g. Sections 6.4.2 and 7.5.7 of ISO 13485 have specific references to sterilization and contamination control).

Our discussion in this chapter will follow the description in the IMDRF QMS document for SaMD, as opposed to that in ISO 13485. This IMDRF document has statements in every section, of the form: "The concepts presented in this section relate to clauses 4 and 5 in ISO 13485" [127]. It also, in addition to being freely available, has the advantage of being software-specific (as opposed to the generic medical device focus of ISO 13485), which makes it a better guide for students.

4.3 The IMDRF QMS Document

4.3.1 Structure of the Document

This document, titled *Software as a Medical Device (SaMD): Application of Quality Management System* [127], provides an excellent introduction to this topic, with a focus on SaMD. (SaMD is defined as "software intended to be used for one or more medical purposes that perform these purposes without being part of a hardware medical device" [125].) The document acknowledges that the software industry has its own good software quality and engineering practices (specifically IEC 62304 [119]), which are used to control software quality. In general, these align well with QMS requirements once allowance is made for patient safety.

Sections 1–4 of the IMDRF QMS are essentially a preamble. Section 5, "SaMD Quality Management Principles," summarizes the core principles of designing/implementing a quality management system for SaMD. Such a system needs to include the following core principles (see also Figure 4.1):

1. an organizational structure that provides leadership, accountability, and governance with adequate resources to assure the safety, effectiveness, and performance of SaMD;
2. a set of SaMD life cycle support processes that are scalable for the size of the organization and are applied consistently across all realization and use processes; and
3. a set of realization and use processes that are scalable for the type of SaMD and the size of the organization; and that takes into account important elements required for assuring the safety, effectiveness, and performance of SaMD.

Essentially, we are moving from an outer circle (Figure 4.1), the organizational structure as a whole, to an organizational culture with defined processes for all of

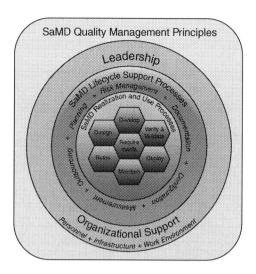

SaMD quality management principles: leadership and organization support, processes, and activities. This illustrates the three layers of a quality management system for developing medical software. Reproduced from figure 1 of the IMDRF QMS document [127].

its products (middle circle), to a set of project-specific processes for the product in question (inner circle). These processes are connected, with each outer layer providing the foundation for the inner layer – see also figure 2 in the IMDRF QMS. Sections 6–8 of that document describe the three layers in detail. We will discuss each in turn next, closely following the structure of the IMDRF QMS.

4.3.2 SaMD Leadership and Organizational Support

Section 6 of the IMDRF QMS is divided into two parts: (1) Leadership and Account-ability and (2) Resource and Infrastructure Management. The leadership of the organization is responsible for establishing and implementing the QMS. Importantly, management needs to ensure both that activities are established for verifying the effectiveness of the QMS (e.g. internal audits) and to ensure that the QMS is sufficient for the task. Key roles need to be explicitly assigned, such as, the person who is responsible for patient safety considerations.

This document uses two fictional companies, "Magna" and "Parva," as examples. This is to illustrate how any given concepts fit a large and a small company, respectively (*Magna* is the Latin word for large and *Parva* the word for small). To quote the specific section:

Both Magna and Parva management have responsibilities to ensure that a QMS has been established and that the necessary patient safety considerations have been built in to the QMS and managed when entering the SaMD market. In the case of Magna, the company has an organizational structure that resulted in its Chief Medical Officer being identified as being responsible for these aspects. In the case of Parva, the company has nominated its Software Development Manager to be responsible for including necessary patient safety aspects.

Note the specificity: the person responsible for patient safety is explicitly named. This is important because unless somebody "owns" the role, this particular task will be ignored. Naturally, busy people focus on the items in their job description, so if something is important, *it needs to be explicitly part of somebody's job description.*[4]

The leadership of the organization is also responsible for proper resource and infrastructure management. In particular, they must ensure that the people assigned to each task are properly trained and have the necessary skills to carry out the software engineering tasks assigned to them. These skills include an understanding of the clinical aspects of the work. The company may need to create internal training courses or provide opportunities for employees to attend external training seminars as needed to acquire such skills/knowledge. The environment in which the work is performed is also important. Management needs to ensure the availability of resources such as computers, testing setup, and server and network infrastructure. This is especially the case if such resources will be obtained from external service providers (e.g. cloud services).

4.3.3   SaMD Life Cycle Support Processes

Section 7 of the IMDRF QMS describes what is needed to ensure an appropriate software engineering culture in the organization. These are common to all SaMD projects undertaken by the company. They include appropriate product planning based on "the quality principle that better results can be achieved by following a methodical and rigorous plan for managing projects such as a plan-do-check-act approach" [127]. The planning phase must take into account both the clinical and the software perspectives. One should also explicitly consider the risk categorization of the proposed product. For more details, see Section 2.2.3 and, in particular, Table 15.1.

Another aspect of planning is risk management, with an emphasis on patient safety – we will have more to say on this in Chapter 5. In addition to schedule and budget risks of a product (which are normal in most software development), in the medical world we have to seriously consider the patient safety aspects arising from both correct use and "reasonably foreseeable" misuse of the software. Risk management is a task that should be explicitly assigned. In large organizations (e.g. hospitals, large companies) there might be a dedicated risk management department, whereas in smaller companies one may need to train senior developers for these tasks.

The next topics address various aspects of "control."[5] The first control aspect is document and record control (or procedures to ensure documents and records), which allows one to provide evidence[6] of what was done and justification for what was not done. As part of the QMS, some of the following activities may be performed to manage and maintain appropriate documentation: (1) reviewing and approving documents before use; (2) ensuring that the current versions of the documents are available to prevent obsolete versions being used; (3) retaining obsolete versions for certain periods; (4) ensuring that no unauthorized changes are made to documents;

and, critically, (5) maintaining and updating the documents as the project progresses. One may use a variety of tools to manage documentation, including source-code control software.

A second related topic is configuration management and control, including source code management, release management,[7] and documentation of the SaMD and its design. In SaMD it is particularly important to ensure the correct installation and integration of the product into the clinical environment. Given that SaMD depends on hardware, part of configuration management is to specify the hardware/network needs of the product to ensure correct installation.

An additional aspect of this part of the QMS is what the IMDRF QMS terms "Measurement, Analysis and Improvements of Products and Processes." This is effectively a self-auditing step to ensure that the processes described are being followed and that the processes used are effective. As part of this, we can then identify ways to improve both the products and the processes. For SaMD, part of this process involves collecting and analyzing quality data, such as, analysis of customer complaints, problem and bug reports, and lack of conformity to product requirements. The IMDRF QMS highlights that for SaMD "customer complaints may be the major source of the quality data that the organization should analyze" [127]. Naturally, all events and all corrective actions should be appropriately documented. Again, explicit responsibility should be assigned to a person/department for managing this task.

The final aspect of this section has to do with outsourcing. One must understand the capabilities of the outsourcing suppliers. One must also ensure clear communication of the roles and responsibilities in such arrangements, the quality requirements that the supplier must adhere to, and the criteria that will be used for the review of the final deliverables. This relates to the discussion of off-the-shelf software in the GPSV [70], discussed in Section 2.2.2. Part of the process is the evaluation of external (potentially open-source) code. It must be emphasized, however, that ultimately responsibility for the safety and performance of the product lies with the core organization, which must ensure that any outsourced work complies with its *internal* QMS requirements. This is easier if the external contractor has a QMS that has been audited/certified as compliant with the QMS of the core organization.

4.3.4 SaMD Realization and Use Processes

The document states that this final section (Section 8) "identifies key life cycle processes that should be identified in the methodologies used in an organization that manufactures SaMD." Here, the term life cycle is an explicit reference to IEC 62304, which defines the life cycle process for medical device software [119]. This essentially follows good software engineering practices. We will have more to say in Part III of this book, which is dedicated to these activities. Software life cycles and QMS are closely related. For example, QMS systems have an explicit focus on following "customer-driven processes." Similarly, the first step in software life cycles is that one must establish user needs and translate these into requirements.

The remainder of this section of the IMDRF QMS follows a standard Waterfall-style software process, proceeding from design (Section 8.2) to development (Section 8.3) to verification and validation (Section 8.4) to deployment (Section 8.5), maintenance (Section 8.6), and decommissioning (Section 8.7). The Waterfall model is simply used to illustrate the concepts. This is primarily due to its relative simplicity. Other life cycle models could also be used. We discuss life cycle models in more detail in Section 9.2 of this book.

In all of these subsections, the document emphasizes that one must account for: (1) patient safety and clinical environment considerations; and (2) technology and systems environment considerations. This recognizes that an SaMD is both a *clinical* and a *software* product, and as such needs to comply with requirements from both universes. Our process has to follow best software engineering practice and explicitly account for patient safety and the use/integration of our software in a clinical environment.

We highlight here a subset of points that are directly relevant from a management/oversight perspective, and revisit in detail the more technical aspects of Section 8 of this document in Part III of this book.

First, we note that for each of the tasks in Sections 8.2–8.7 of the IMDRF QMS document, one needs to have an explicit plan/procedure in the organization's QMS, with explicitly assigned personnel that will be responsible for carrying it out.

A second point worth highlighting is the statement (Section 8.2 of the IMDRF QMS document) that "building quality into SaMD requires that safety and security should be evaluated within each phase of the product life cycle and at key milestones. Security threats and their potential effect on patient safety should be considered as possible actors on the system in all SaMD life cycle activities." Safety (exposure of the patient or the user to any undue risk from use of the software) and security (exposure of patients and users to undue risk from unintentional and malicious misuse) must be explicitly accounted for at the design stage and kept in mind throughout all phases of the process. In addition to more obvious issues such as physical harm to the patient, in this internet era one must ensure the use of appropriate safeguards for patient privacy with respect to data storage and management. To further emphasize this point, the very next statement in the document explicitly pairs patient safety and confidentiality.

Section 8.3 of the document states that the "implementation of clinical algorithms adopted should be transparent to the user in order to avoid misuse or unintended use." This need for transparency could have interesting implications for the use of black-box deep learning techniques that are becoming increasingly popular.

We need to highlight one final important point. Often, an SaMD is part of a pipeline/workflow/clinical process and thus needs to integrate with other systems (Section 8.5 of the document). This is not a simple "download and install" process, but one that requires integration and collaboration with hospital IT, hospital risk management, and other entities. As the document explicitly states, "there should be communication of relevant information to enable correct installation and configuration of the SaMD for appropriate integration with clinical workflows. This can include instructions on how to verify the appropriateness of the installation and update to SaMD as well as any changes made to the system environment."

The IMDRF QMS is a relatively short, freely available document, and is well worth reading in its entirety.

4.4 Implications for Research and Scientific Software

To reiterate a point made multiple times earlier in this book (e.g. see Section 1.1.2), it is the common experience of experts and regulators that an organization/company that is producing a medical device (or software) must be organized and structured in such a way as to ensure product quality – hence the term "quality management." Unless the software is designed and implemented within such a "quality culture," it is unlikely to be of high quality and as such could not (should not) be trusted in situations that entail risk to patients and users. This is often contrary to how many junior software engineers think.[8] The quality processes (documentation, reviews, etc.) exist to create an environment that will allow a high-quality piece of software to be developed. They are not obstacles to be worked around (however annoying they appear), but the necessary guardrails to ensure the success of the final product.

Unfortunately, this is a lesson that is not well understood by many outside regulated industries, and especially in academic research. For example, many highly used scientific software packages in labs emerge organically from the bottom-up with no organizational support/oversight. Many such software packages become popular and eventually play a critical role in our understanding of many scientific processes. In functional brain imaging, for example, even large, commonly used software packages[9] contain errors (bugs) that may go unnoticed for years [57, 266]). While software errors will always exist, software that guides future treatments and impacts human disease, even at the research level, needs to be designed for the highest level of quality. Flaws in such software can invalidate otherwise good research and lessen the impact of the work of the scientific community as a whole.

The same lessons apply to proper quality management of the processes used in the training and validation of machine learning (ML) algorithms. This type of work often results in high-impact publications that steer the future development of these technologies. However, we suspect that the results of some of these papers are compromised by either data contamination (i.e. the accidental mixing of training and testing data), by testing and training on overly homogeneous datasets, or by training far too many models (or variations of the same model) on the same dataset, a process analogous to p-hacking [107]. This type of problem can artificially inflate the performance metrics of ML algorithms. Early regulatory guidance suggests using completely different sources

neil_ferguson ✓
@neil_ferguson

I'm conscious that lots of people would like to see and run the pandemic simulation code we are using to model control measures against COVID-19. To explain the background - I wrote the code (thousands of lines of undocumented C) 13+ years ago to model flu pandemics...

5:13 PM · Mar 22, 2020 · Twitter for iPhone

Figure 4.2   Imperial College London pandemic simulation code. This tweet from Professor Neil Ferguson (Imperial College, London, UK) accidentally summarizes many of the problems with "home-grown" lab software being used for mission-critical tasks – in this case the management of COVID-19 in the UK and beyond. No QMS-based organization would have allowed the incorporation of thousands of lines of undocumented code in a product, and no regulator would have ever cleared it, even in a low-risk scenario.

of data for validation purposes than what was used for training, and even using an external organization to perform the validation work – this is discussed in Section 15.1.2.

Similar lack of quality management issues are evident in software used to model the spread of pandemics. A recent example of a package that was used to make very consequential decisions is discussed in Figure 4.2.

4.5   General Remarks

In many of the case studies that we will review in Part IV of this book, things went badly wrong. While some of the issues could simply be attributed to human error, the background reality of many of these is a failure to follow a proper process, often a result of organizational and leadership failure. A common thread in these is deadline pressure to launch a product.[10] In others, there was an attempt to take shortcuts (e.g. the Boeing 737 MAX disaster – see the vignette presented in Chapter 21 – and the Iowa Caucus app – see the vignette presented in Chapter 20) to reduce costs and regulatory oversight.

It cannot be emphasized enough that the common wisdom of both practitioners and regulators seems to place an increasing emphasis on organizational culture. To summarize: A well-managed organization is necessary[11] in creating a high-quality, safe product. This notion is reinforced by the FDA as part of its pilot digital health software precertification pilot program [85], which reduces oversight on well-run organizations.

4.6   Conclusions

A critical regulatory requirement for the development of medical software is that a company must operate under a QMS. The quality system governs all aspects of the process, beginning with how the organization functions in general, to how it develops software (e.g. policies for source-code management), to how the specific software project under consideration is organized. Both regulators (such as the FDA) and the international standards are placing an increasing emphasis on the operation of the organization as a whole, to the extent that we are seeing programs, such as the FDA's precertification program, where the FDA's regulatory review process is mostly focused on the organization and not on the actual product.

4.7   Summary

- A medical software/device manufacturer must have a QMS in place that governs all aspects of its operation.
- The QMS affects not only how software is developed and maintained, but also how the company, as a whole, is run.
- The experience of regulators is that unless software is designed and implemented within such a "quality culture," it is unlikely to be of high quality and as such could not (should not) be trusted in situations that entail risk to patients and users.
- The lack of a QMS in many research enterprises (e.g. universities) raises serious questions about the quality of software produced as part of such research and how reliable such software is when used for more widely.

RECOMMENDED READING AND RESOURCES

The main international standards in this area topic are:

ISO:9001 Quality management systems, 5th ed., 2015.

ISO:13485 Medical devices – quality management systems – requirements for regulatory purposes, 2016.

ISO/IEC/IEEE:90003 Software engineering – guidelines for the application of ISO 9001:2015 to computer software, 2018.

A good example of regulatory guidance for QMS is the following IMDRF document that we used as a basis for much of the discussion in this chapter:

IMDRF, SaMD Working Group. Software as a medical device (SaMD): application of quality management system, October 2, 2015.

Finally, there are a number of books that act as a guide to the ISO 9001 standard. Two good examples are:

A. Jarvis and P. Palmes. *ISO 9001:2015: Understand, Implement, Succeed!* Addison-Wesley, 2015.

C. Cochran. *ISO 9001:2015 in Plain English*. Paton Professional, 2015.

NOTES

1. A reminder for the reader: unlike guidance documents, this is an official US government regulation and as such has the force of law in the United States. There have been cases of individuals being prosecuted for violating their own QMS!

2. In a world where AI/ML methods are becoming more common, this process should include procedures for training data management, as discussed in Section 9.3.1.

3. This raises some potential concerns about both regulatory capture [151] and companies becoming complacent as their scrutiny is relaxed.

4. As the common aphorism states: "Everybody was sure that Somebody would do it. Anybody could have done it, but Nobody did it."

5. If the reader finds the term "control" confusing, it might be helpful to replace it with the phrase "procedures to ensure." See also the FDA's guidance on Design Controls [88].

6. In all regulated activities, as mentioned before, the word "evidence" should be read as "evidence provided in a legal setting," – that is, a trial hearing in a law court.

7. To make an obvious point here: Unlike hardware, in which the product is a physical device that requires a factory facility to build, software can be easily changed and recreated by a developer with access to the source code and an appropriate development environment. One must, therefore, be careful to ensure that the released version consists of the correct set of files, and that it has not been tampered with.

8. As one of the reviewers of this book pointed out, this is not to imply that programmers should not experiment to find the right solution. Such experimentation, however, needs to be carried out outside of the product, and when the right solution is found it can be used to define the specification and tests so that this solution can be introduced correctly and efficiently within the procedures of the QMS. It is also worth noting that quality assurance (QA) personnel must recognize that the correct implementation is usually not "obvious" and that these experiments need to take place and, therefore, must be facilitated by the organization.

9. One naturally must acknowledge that many of these packages are developed on a shoestring budget, often by a small team of people or even a single individual. Some of these developers are extremely gifted and often produce excellent/reliable code even in the absence of formal quality management. However, this is not a generalizable strategy.

10. Sometimes literally "launch" – see the description of the Mars Climate Orbiter disaster in the vignette in Chapter 18.

11. Necessary but not sufficient! One has to have all the other attributes, such as technical competence and appropriate resources.

5 Risk Management

INTRODUCTION

This chapter presents an introduction to risk management, a core regulatory requirement for all medical software. We begin with an overview of the regulatory background (Section 5.1) and then review both the international standard ISO 14971:2019 and a recent guidance document from the IMDRF. Next, we describe the process of risk analysis (Section 5.2), including issues related to the use of artificial intelligence/machine learning (AI/ML) techniques. We then discuss risk evaluation (Section 5.3) and risk control strategies (Section 5.4). The chapter concludes with a section (Section 5.5) that briefly outline the role of risk management in the software life cycle process described in Part III, where risk management will be our ever-present companion.

5.1   An Overview of Risk Management

5.1.1   Regulatory Background

When designing medical software, one must always be mindful of the potential dangers to patient safety. This is not simply a "nice thing," but a clear expectation from the regulations and international standards governing medical software. Safety is prominently listed in the FDA's mission statement (see Section 2.1), with its focus on "safety, efficacy and security." Risk management is a critically important tool in ensuring patient safety.

The main international standard on this topic, ISO 14971, defines risk management as the "systematic application of management policies, procedures and practices to the tasks of analyzing, evaluating, controlling and monitoring risk" [133]. To understand this definition, we need to define three key terms (again following ISO 14971):

- **Harm:** Injury or damage to the health of people, or damage to property or the environment.
- **Hazard:** A potential source of harm.
- **Hazardous situation:** A circumstance in which people, property, or the environment are exposed to one or more hazard(s).
- **Risk:** The combination of the probability of occurrence of harm and the severity of that harm.

The event chain is as follows: Despite our best efforts in designing our software, there is the possibility of things going wrong. These are the *hazards*. In certain circumstances, such a hazard, if not identified in time, can lead to a *hazardous situation*, a circumstance in which the patient is exposed to the effects of the hazard. The effects of the hazard can then lead to actual *harm* to the patient (or a caregiver). The *risk* from a specific harm is a combination of how likely the harm is and how serious the harm is. For example, a situation that may result in patient death is always a high-risk situation, regardless of the probability of its occurring.

How do we manage risk? Before discussing this we define some additional terms [133]:

- **Risk analysis:** The systematic use of available information to identify hazards and to estimate risk.
- **Risk control:** The process in which decisions are made and measures implemented by which risks are reduced to, or maintained within, specified levels.
- **Residual risk:** The risk remaining after risk control measures have been implemented.
- **Risk management:** The systematic application of management policies, procedures, and practices to the tasks of analyzing, evaluating, controlling, and monitoring risk.

When creating a medical device or medical software, one performs a risk analysis to identify hazards and to estimate the risk. Risk control measures are then taken to manage the risks and ideally reduce their probability, their severity, or both. The remaining risks are the residual risks, which must be within some acceptable bounds. In validating a medical device (ensuring that it meets the user's needs), one must take risk into account. As stated in the FDA's Quality System Regulations, "Design validation shall include software validation and risk analysis, where appropriate. The results of the design validation, including identification of the design, method(s), the date, and the individual(s) performing the validation, shall be documented" [69].

Much like every other process in medical software design governed by the overall quality system (see Chapter 4), risk management will involve the creation of a specific risk management document (plan) that specifies the appropriate procedures to be followed. Key to the formulation of this plan is the identification of the risk category into which our proposed software (or device) fits. In Section 2.2.3, we reproduce a table (here Table 2.1) from an IMDRF document [126] that provides some more details as to how to identify the risk category into which our software fits. This is the first and critical step in designing a risk management plan.

Please note that our focus in this chapter is exclusively safety/security risk – that is, risk of harm to a patient, a caregiver, or somebody else present nearby. There are other risks possible in a project, including cost risk (the project is too expensive), technical risk (we fail to create the product), schedule risk (the project takes too long), organizational risk (high-level structural issues with management impede the project), and market risk (nobody buys our software). The treatment of these risks is, in principle, similar and can be accounted for appropriately in risk management plans. See Section 1.2 and Figure 1.2 for a brief discussion of these other types of risk.

In particular, it is important to consider risk management within the overall perspective of business strategy and law. Specifically, the business analysis would be "what's the potential downside?" which may be measured in expected legal fees and insurance premiums associated with the adverse event (here, fees could be lawyer costs, settlement, or judgment). For big companies, which are often self-insured, this is a real cost that may happen at any time. For smaller companies, risk could translate into the cost of insurance against such an event. Hence, risk analysis and management is not simply a regulatory exercise but also directly impacts the operational costs (and thus the bottom line) of a company.

5.1.2 The ISO 14971:2019 Standard

This recently revised standard has a total of 49 pages. The outline of the document is a helpful indicator of what it covers. After the usual introductory sections (scope, references, terms, and definitions), we have:

- Section 4 – General Requirements for Risk Management Systems (3.5 pages);
- Section 5 – Risk Analysis (2.5 pages);
- Section 6 – Risk Evaluation (1/3 page);
- Section 7 – Risk Control (2 pages);
- Section 8 – Evaluation of Overall Residual Risk (1/3 page);
- Section 9 – Risk Management Review (1/3 page);
- Section 10 – Production and Post-Production Activities (1.5 pages);
- Annex A – Rationale for Requirements (8 pages);
- Annex B – Risk Management Process for Medical Devices (4 pages);
- Annex C – Fundamental Risk Concepts (6 pages).

The annexes are explicitly marked as informative – that is, they do not contain requirements that a company that tries to comply with the standard must follow. They are, however, very valuable, as they provide useful explanations and examples. We will follow much of the discussion in these annexes in this chapter.

This is the third edition of this standard. The first edition was published in 2000. The motivation for publishing this standard (see Annex A of ISO 14971 [133]) was a recognition that "absolute safety" in medical devices was not achievable and that the risks that come with the increasing diversity of medical devices and their applications could not be managed through safety standards. There was therefore a shift from absolute safety to risk management, which led to the publication of the first edition. The second edition (2017) provided additional guidance on the application of the standard, and in particular the relationship between *hazards* and *hazardous situations*. This last edition (2019) clarifies the requirements, and in particular describes in more detail the concept of *residual risk*.

5.1.3 General Principles of Risk Management

We will follow here the discussion in Section 7.2 of the IMDRF QMS [127]. As a reminder, this is a document issued by the International Medical Device Regulators Forum (IMDRF). This section is very short (1.5 pages) and should be read in its entirety by anybody involved in this type of work. Here are some salient points. The first is that the risk management process should be integrated into the entire life cycle of an SaMD (software as a medical device). Risk should be considered at all steps (design, implementation, verification, validation, deployment, maintenance, and retirement) of this process – see Section 4.3.4. It is not a one-off activity that is performed at one stage of the process.

When evaluating the risk arising from the use of an SaMD, we need to account for both what might happen during the normal use of the software and (critically) "reasonably foreseeable misuse." We need to understand the environment in which our software will be used and the issues that might arise from this. Finally, specific to software-only devices, the relative ease of updating, duplication, and distribution create additional risks that need to be accounted for. According to the IMDRF QMS, in the company's quality management system one should include [127]:

- identification of hazards;
- estimation and evaluation of associated risks;
- actions to control risks; and
- methods to monitor effectiveness of the actions implemented to control risks.

In identifying hazards, the IMDRF QMS proceeds to helpfully identify five broad sources for these:

1. User: These are particularly relevant for software that is expected to be used by a patient in the home environment.
2. Application: Is there a need to restrict availability of the software on certain devices to mitigate user risk?
3. Device: Here, the particular danger is small devices with small screens not providing all the information or making it hard to use.
4. Environment: Risks created by environmental disruptions (e.g. loss of network connectivity, background noise, bad lighting).
5. Security: In a connected world we need to worry about security threats and include features such as intrusion detection, data integrity, and loss of data. (Cybersecurity, which is discussed in Section 3.5, must also be accounted for in the risk management process.)

One final note from the IMDRF QMS has to do with the balance between security (issues arising from malicious or unintentional use) and safety (issues arising from the intended use) risks. The recommendation from the IMDRF is that one should ensure that security controls do not take precedence over safety considerations.

In the next section, we discuss the risk management process in more detail. ISO 14971 [133] divides this into risk analysis, risk evaluation, risk control, and production and post-production information.

5.2   Risk Analysis

Risk analysis is the process of systematically using all information available to us about our product, its users, the environment, and any other factors to identify hazards and the risk associated with these hazards. In our discussion, we will follow the structure of Section 5 of ISO 14971 [133].

The first step in risk analysis (see Section 5.2 of ISO 14971) is to list the intended uses and reasonably foreseeable misuses (incorrect or improper uses) of our software. The uses part is straightforward, as the uses of our software are part of the design and should be found in the system requirements – see Chapter 12. Misuses relate to how our software could be used improperly. These include, for example, common errors or using the software for tasks for which it was not meant to be used.[1]

The next step is the identification of characteristics related to safety (Section 5.3 of ISO 14971). This is probably more relevant to hardware rather than software and has to with degradation.

5.2.1   Hazard Identification

We follow here Section 5.4 of ISO 14971. In the beginning of a software project, hazard identification is essentially a predictive process (see also Vogel [261, p. 118]). This process depends on the overall experience of the team (including clinical and legal specialists[2]), largely based on prior history with this type of software by the team itself or by other companies/competitors. For example, in medical imaging a well-known risk is accidental left–right flipping of an image, which can result in (in an extreme case) surgeries being performed on the wrong side of the body. Any experienced imaging specialist will list this as a concern and ensure that the software has appropriate checks for this type of error.[3]

As the software process progresses, one may also obtain a sense of what the hazards are from issues arising from the behavior of the software (e.g. faulty behavior under heavy load) or even more critically from tester/customer reports and complaints.

It is also important to encourage all participants to try to think outside the box (in a brainstorming fashion) as to what might constitute a hazard and to especially encourage more junior members of a team to speak up if they see something that everybody appears to be missing. As General Patton is reputed to have said: "If everybody is thinking alike, then somebody isn't thinking."[4] This is particularly applicable when evaluating potential cybersecurity threats, where one needs to think more like a hacker/criminal and less like an engineer.

5.2.2   Estimation of Risk

The next section of ISO 14971 (Section 5.5) addresses the problem of risk estimation. As a reminder, risk is a combination of probability and severity. The estimation of risk therefore involves estimating the probability of something happening[5] and the severity of the consequences to the patient/user. Estimating probabilities for software is not a trivial matter, however, as most software is completely deterministic – that is, it produces the same outputs given the same inputs.[6] With this in mind, what is often perceived as random software failure is really a bug that only manifests itself when a certain set of inputs (input history, really) is given to the software. This is unlike hardware, where, for example, micro-structural defects can introduce a random variation in the expected lifetime of a component.

In examining software for likely points of failure, one can use knowledge of the different components (e.g. complexity, history of failure, experience level of the developers, changes late in the history of the project, quality of testing, possibility of user confusion) to estimate the likelihood of failure in any of them. Certainly a complex component that was implemented by an inexperienced developer and that has not been fully tested is highly likely to fail!

Severity is in many ways easier to assess. To assign the level of severity, we can assume that the harm has occurred and then proceed to estimate such a harm's consequences for the patient/user. For both severity and probability, it may also be better here to use a discrete categorization scale (e.g. in the case of severity one can use categories such as negligible, moderate, serious, life threatening) to characterize these events.

5.2.3 A Risk Estimation Example

In this section, we follow the discussion in Annex C of the ISO 14971 [133] standard. The orienting flowchart is shown in Figure 5.1. This is best described by means of an example. We adapt here one of the cases in Table C.3 of the standard. Consider the case of a software package whose goal is to measure the size of a brain tumor from magnetic resonance (MR) images. The diagnosis and subsequent patient management are critically dependent on the size of the tumor, so this is a very important measurement.

1. A potential *hazard* is that our software produces a bad measurement.
2. The *foreseeable sequence of events* is (a) there was a measurement error, and (b) the user (radiologist) failed to detect this error.
3. The *hazardous situation* is that this error can lead to a bad diagnosis and ultimately a failure to administer appropriate therapy.
4. The resulting *harm* is that diagnostic and/or therapeutic failure can lead to a worse clinical outcome for the patient and potentially death.

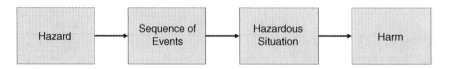

Figure 5.1 From hazard to harm. This flowchart illustrates the relationship between hazard, hazardous situation, and harm. Risk is defined as a combination of the probability of harm occurring and the severity of harm. We discuss these topics in more detail in the main text. Figure adapted from figure C.1 in ISO 14971 [133].

With this flow in mind, how does one estimate risk? Risk is a combination of the probability of harm and the severity of harm. The first is to try to compute the probability of harm.[7]

The first step is to estimate the probability that a hazardous situation will occur, given the sequence of events. We can define p_e as the probability of error and p_d as the probability of failure to detect the error. Assuming that the events are independent, the probability of arriving at a hazardous situation is $p_{HS} = p_e \times p_d$. The next step is to estimate the probability of the harm occurring. Not all hazardous situations lead to harm. In our particular example, the wrong measurement may not affect the overall treatment plan if it is only one part of the decision process that leads to this, which has probability p_B. Finally, the overall probability of harm is the product $p_{HS} \times p_B$. If, for example, the probabilities are $p_e = 0.1$, $p_d = 0.2$, and $p_B = 0.5$, we arrive at a combined probability of 0.01.[8] At once, of course, one realizes that computing such probabilities is far from trivial. We will revisit this topic later in this chapter.

The second step is the quantification of severity. This is probably easier, as we know what the consequences of the harm to the patient could be.

Finally, risk is the combination of probability and severity. Risk is higher when we have both high probability and high severity. It can be made less "dangerous" when either of these factors is reduced. While one can pretend to be "scientific" and assign probabilities to multiple digits of precision, the reality is that it is often better to use a qualitative scheme and assign values such as low, medium, and high for both probability and severity. Such gradations are easier to understand (especially if the situation is time-critical), and the coarseness is probably a more honest reflection of our confidence in assigning these. Following this process we may end up with a risk categorization scheme similar to the one shown in the table shown in Table 5.1.

The temptation to set the probabilities to lower values than is warranted should be immediately recognized – see the vignette presented in Chapter 21 for an example. In this case, Boeing "overestimated pilot skill while underestimating the probability for software errors" [56].

Table 5.1 Risk categorization level. This table shows a *qualitative* risk categorization scheme adapted from Vogel [261].

Risk level	Probability of harm	Severity of harm
1	Low	Low
2	Low	Medium
2	Medium	Low
3	High	Low
3	Low	High
3	Medium	Medium
4	Medium	High
4	High	Medium
5	High	High

5.2.4   Risk Analysis for AI/ML Modules

The black box nature of AI/ML modules introduces the potential for additional risk. This is particularly the case if the module does not provide any additional information other than its output (see also the short discussion on comprehensibility in Section 9.3) that permits a user to assess whether the output is correct.

In the document DIN SPEC 92001-1:2019, produced by the German Institute for Standardization, there is a description of the need to evaluate AI/ML modules for robustness as a key measure of quality. In particular, there is a concern about "the ability of an AI module to cope with erroneous, noisy, unknown, and adversarial input data" [53]. The complexity of a module that uses high-dimensional, nonlinear functions means that even small changes in the input can result in large changes in the output. This results in the need to explicitly assess the robustness of such modules to small changes in the input. The document describes two potential sources of error:

1. **Distributional shifts**: This is when the module is given inputs that are outside the training and testing data sets.
2. **Adversarial attacks**: This happens when someone purposely provides slightly altered data for malevolent purposes [6, 199, 244].

The document goes on to emphasize the need to directly address the possibility of adversarial attacks, as these can pose major safety and security risks, and potentially implement defensive and monitoring strategies to reduce the risk involved. These involve having a proper cybersecurity process (see Section 3.5).

5.3   Risk Evaluation

ISO 14971 [133] has a very brief section (Section 6) that discusses risk evaluation. The process can be summarized as below:

For each hazardous situation that we have identified in our risk analysis:

- evaluate the estimated risks;
- determine if the risk is acceptable or not:
  - If the risk is acceptable, document and stop.
  - If the risk is not acceptable, perform risk control activities to reduce the risk.
  - At the end of risk control, what is left is the *residual risk*, which is evaluated in the same way.

The definition of acceptable is dependent on the project and its categorization (see Table 2.1). It is a management decision subject to regulatory guidance. For example, if we consider Table 5.1, which mapped risk on an 1–5 scale, it may be that for our project the acceptable level is 2. Then, for each hazardous situation, we compute the risk. If this is less than or equal to our acceptable risk level (2 in this case), our task is completed. If not, we follow appropriate risk control activities (see the next section), and at the end of these activities we compute the residual risk – the risk after risk control. If it is greater than the acceptable level, we must repeat the risk control process to try to improve things. Should this process fail to reduce the residual risk to acceptable levels, we may need to perform substantial changes to our design to eliminate the hazard.

The level of acceptable risk depends on the situation and often results from a risk–benefit analysis. There are many high-risk operations that are deemed acceptable if the potential benefit to the patient is great. As a generic example, experimental drugs can be given to terminally ill patients. While the risk of death might be substantial, for patients in this situation the benefit (e.g. 20 percent probability of cure) might outweigh the risk of death, given that in the absence of the experimental drug their life expectancy is limited.

5.4   Risk Control

To repeat the definition from ISO 14971 [133], *risk control* is the process in which decisions are made and measures implemented by which risks are reduced to, or maintained within, specified levels. In our discussion we will follow the description presented in Section 7 of this standard.

5.4.1   Risk Control Strategies

For each hazard where the risk is unacceptable, the guidance in the standard is to determine which of the following measures is appropriate for reducing the risk to an acceptable level. There are three basic strategies:

1. Change the design to eliminate the possibility of the hazard occurring.
2. Add preventive measures in the software to eliminate or reduce the probability of harm.
3. Add safety information measures (e.g. warnings to the user, extra training).

The guidance from ISO 14971 is that these measures should be used in the order listed above. The implication here is that safety information measures should be used as a last resort if the other options are not available. The preferred method is, naturally, eliminating the probability of hazard occurring altogether.

The first and preferred strategy is one in which we change our design to eliminate the problem. Consider the case of software that can, in addition to other functions, automate a certain surgical procedure using fully automated robotic control. If the risk level in such a procedure is too high and the risk cannot be eliminated, we may choose to remove the functionality from the software to eliminate the risk altogether. This is a surprisingly common and effective strategy.

The second strategy involves adding checks to the software to eliminate or reduce the probability of harm. In the case of left–right flipping in medical images, one (granted, a hardware) solution is to attach a small, bright capsule to the right side of the head during imaging. Then, one adds as part of the procedure the identification of this capsule in the image. This will allow the user to accurately verify the left–right orientation of the image.

The final strategy is less satisfying and involves providing additional information for safety and potential extra training to the users so that they can recognize hazardous situations when they occur. This may involve changes to the software to issue warning messages in those situations where we can detect that the generated result from our algorithm is likely to be wrong. This may warn the user to check the process again. The problem with this type of approach is that it moves the problem forward to the user, and the user may ignore the warnings – see the case of "Malfunction 54," for example, in the vignette presented in Chapter 17 that describes the Therac-25 accidents. The technologist ignored this message (in fact did not even know what the message meant as it was not documented), which led to the delivery of a lethal dose of radiation to the patient.

5.4.2   Post Risk Control Evaluation

Once the control strategies are implemented, the risk analysis process is repeated to ensure that the residual risk is acceptable. If it is not, additional control measures may be needed to reduce the risk further. Sometimes, we may find ourselves at a dead end. It may well be that we have a situation where the risk is not acceptable according to our criteria, and that additional risk control methods are not possible/practicable. In this case, the manufacturer has the option of gathering additional data to determine whether the benefits of the intended use (the use that results in the hazard) outweigh the residual risk. If this is the case, we effectively consider the residual risk acceptable. If, on the other hand, the risk is seen to outweigh the benefit, then the risk is *unacceptable*, and we need to consider modifications to the software (and/or its intended uses) to resolve the problem.

5.5   Risk Management and the Software Life Cycle

In Part III of this book, we present a recipe for creating medical software. This is subdivided into chapters, each focusing on a different aspect of the process. The astute reader will recognize at once that there is no "Risk Management" chapter in this part of the book. There are, rather, "Risk Management" sections in each chapter. This is because risk management is an ongoing activity. Hazards can arise from the entire process, including requirements, design, implementation, testing, and all other steps. Risk management, therefore, is an ongoing activity that needs to be revisited at each step of the process. Figure 10.2, which is adapted from the IEC 62304 standard [119] on software life cycles, graphically illustrates this point as well. One can see that the box labeled "Risk Management" extends throughout the timeline of the process.

> "Everything we do before a pandemic will seem alarmist.
> Everything we do after will see inadequate."
>
> *Michael Leavitt, Secretary of Health and Human Services 2007*

Figure 5.2 A prophetic quote from 2007. Risk management can often be seen as alarmist and expensive. Anyone proposing to spend significant amounts of money on pandemic preparation in 2007 would have been seriously questioned.

5.6 Conclusions

Risk management is an essential, mandated (by regulations) component of the medical software process. It is the process of evaluating every component of every step of the software life cycle (from design to release) for the possibility of causing harm to a patient (or user), assessing the severity of the risk involved and taking appropriate measures to reduce this to acceptable levels. When risk management is done correctly and our software does not exhibit any of the feared behaviors, it will be tempting to think that the whole process was unnecessary and alarmist. This is similar to software testing. Both are "negative" activities and have the goal of preventing "bad" things from happening, as opposed to enabling interesting/useful functionality. These negative activities do not help with the marketing of the software in the short term, though in the long term they help create a reputation for quality.

An example of a successful risk management process was the handling of the so-called "Y2K" crisis. This was caused by a significant fear of computer failure as the year 2000 dawned due to old software that used two-digit numbers to store the calendar year. Extensive risk control work was done prior to December 31, 1999, ensuring that there were no serious problems. We discuss this in the vignette presented in Chapter 22.

A second example is illustrated by the quote shown in Figure 5.2. This is a statement made in 2007 by Michael Leavitt, who was the US Secretary of Health and Human Services. The quote summarizes many of the issues with the perceived lack of value of negative activities such as risk management.[9] When there are no disasters (or before a disaster), such activities appear alarmist and unnecessary. After a disaster occurs, nothing we do will feel adequate. As the old adage goes: "better safe than sorry."

5.7 Summary

- Risk management is a necessary activity in the development of any medical device (and software). This is explicitly mandated in the regulatory documents.
- Risk is the combination of the probability of the occurrence of harm (injury or damage) and severity of that harm.

- Risk management consists of estimating the risks that use of our software could create (including reasonably foreseeable misuse) and taking measures to change the software (or provide additional information/warnings to the user) to reduce either the probability or the severity of these risks, should they be assessed to be unacceptably high.
- Risk management is an activity that is performed at every stage of a software life cycle (design, implementation, testing, validation, deployment, maintenance, retirement).

RECOMMENDED READING AND RESOURCES

The definitive guide to risk management in medical devices is the international standard:

ISO:14971:2019 Medical devices – application of risk management to medical devices, 2019.

The IMDRF has two regulatory guidance documents that are also worth consulting:

IMDRF, SaMD Working Group. Software as a medical device (SaMD): application of quality management system, October 2, 2015.

IMDRF, SaMD Working Group. Software as medical device: possible framework for risk categorization and corresponding considerations, September 18, 2014.

NOTES

1. For example, a particular software could be designed to analyze a particular type of MR image. If the software is given a different type of MR image, perhaps from a new scanner, this could result in a hazard caused by misuse of the software.
2. Any past history that resulted in expensive legal proceedings/significant injuries will play a significant role here.
3. The head/brain area, in particular, is largely left/right symmetric, so it is impossible for a person to identify this problem visually. This is less of a problem in the thoracic and abdominal regions, where there are large asymmetries.
4. George S. Patton Jr (1885–1945) was an American general during the Second World War.
5. This is typically a lot easier to do in hardware devices as opposed to software. Consider the case of a surgical stapler. One can sample from a batch of staplers made during the same time period and test these multiple times to measure the failure rate. Doing this for software is much, much harder.
6. There are some exceptions to this fully deterministic picture. One example is the use of random seeds to initialize the optimization algorithms used for training deep neural networks. However, these random effects should only result in insignificantly small changes in the outputs.
7. For an introduction to probability, see Section 8.2. For the present discussion, it suffices to define the probability of an event occurring as a number between 0 (will not happen)

and 1 (certain). We also define the probability of two independent events occurring, where event 1 has probability p_1 and event 2 has probability p_2, as the product $p_1 \times p_2$.

8. One lesson from this calculation is that the longer the event chain, the safer the system. If, for example, our system had two people needing to check the measurement, each with a probability of error of 0.2, then the overall probability of harm would be reduced to $0.2 \times 0.2 = 0.04$. This is a practical illustration of how a system of checks and balances reduces risk. Having many independent checks reduces the probability of something going wrong.

9. These lines are being written in the middle of the COVID-19 pandemic.

6   Taking an Idea to Market: Understanding the Product Journey

INTRODUCTION

This chapter presents the "business" view of medical software. It takes a company to bring an idea to market and, ultimately, clinical use. We begin with a brief description of issues related to entrepreneurship (Section 6.1): Should somebody who has a promising idea consider starting a company on their own? Next, we discuss the issues of user-centered design (Section 6.2) and of articulating the value proposition of a new product (Section 6.3). We then take a detour through the minefield that is ensuring proper intellectual property protections (Section 6.4). Finally, we discuss the process of raising capital to support a new venture (Section 6.5). While the material in this chapter is focused on startups, it is also important for those that are employed in large companies, as this material will give them an understanding of what their managers are worrying about. In a large company, the capital-raising and marketing may happen internally as a particular group within a company tries to obtain support for a new product from higher-level management. The process looks different, but the fundamental concepts are just as applicable.

6.1   Entrepreneurship: Should I Start a Company?

There have been many efforts to bend the cost curve in healthcare, and with the advent of connected health technologies we are certainly seeing an accelerated shift toward patients accessing their data faster and driving the next steps in their healthcare. Recent legislation in the United States in fact allows patients to view the results of imaging or laboratory testing at the same time the clinician does within the electronic health record (EHR) portal, making the need for integration of healthcare IT even more critical. In the United States and Europe there is an increased demand for software products that allow interoperability between EHRs and outside products, and for an entrepreneur this creates a wealth of opportunity in the chaos. Our recommendation for whether or not an early-stage software product or medical device should result in the formation of a business enterprise may surprise you. The single most fundamental question that we ask all students in healthcare-venture classes or in early-stage ventures is the following:

Do you have a customer who is willing to pay cash for your product? Whether the price point is one dollars, one euro, one million dollars, or one million euros, the mere fact that someone in the healthcare workflow process believes this product will improve their efficiency, provide greater accuracy, or, best of all, provide a solution to a "burning problem" they have faced for years is the single best reason to start a company.

Perhaps a tangible example will help elucidate this point. Several years ago, a few young entrepreneurs wanted to ascertain why it is so difficult for patients in the United States to know when to take their medications and which pills were to be taken at what time of the day. A person on six or seven different pills a day may easily get confused, especially as prescriptions change. Using human-centered design, this savvy group of entrepreneurs sat and watched the entire process from how pills are dispensed to how people fill and use "pillboxes" in their homes, only to discover with horror that errors occurred along each step of the process. If the customer, the patient in this case, could be provided with an easier and more enjoyable experience with less stress, and the technology could provide the buyer (the insurance company in this scenario) with a more accurate alternative to in-person pharmacy pickup it would be a win–win situation. Using human-centered design, the founders of Pillpack™ figured this out beautifully. Pillpack did not start as a company until they had spent months evaluating customer needs, mapping the user journey, and assessing the market to see who might pay for the product [242]. Once they knew there were a few possible customers and that the current patient experience had drastic room for improvement, they launched as a company.

The second key lesson, as shown in this illustrative case, is that their company had to allow room to pivot several times from the elder medication market to the herbal market to the mail pharmacy alternative that worked with insurance plans. If you wait to launch a company until you have already decided all the revenue streams as well as the workflow and have lost the nimble flexibility to pivot, it will quickly become evident how difficult it is to launch a medical software or device company [15]. Regardless of the caveats mentioned here, many in the healthcare and investment industry feel this is an exciting time in healthcare entrepreneurship – with an opportunity to bend the cost curve, given the ability of technology to create a global healthcare conversation.

Crisis and chaos within a healthcare climate often create the need for new technologies and products that have been available before but have not seen the same market penetration as they begin to when the external environment changes. During the time of the COVID-19 crisis there was an increased need for safe methods of provider–patient interaction, resulting in several telehealth and digital health solutions in the marketplace. This example of the pandemic shifting the way healthcare is delivered so dramatically, and potentially for the long term, highlights the opportunity created by chaos in the marketplace. When an entrepreneur asks themselves, "Should I start a company?," certainly anyone with a health IT software solution and the ability to provide telehealth services may have benefited during COVID-19, given the external climate.

However, once a given crisis passes, the urgency to acquire certain products (e.g. telehealth) is reduced. Post-crisis defects in the design or ease of use of a software interface may not be as readily overlooked by either investors or large companies acquiring a startup as they would have been during the pandemic when the urgency of need overshadows detailed evaluation – thereby better explaining the value of timing, often referred to by entrepreneurs as "luck". [144]. One method to ensure that the software product has a long-lasting impact that survives the acute healthcare climate is to really pay attention to designing for the user, with a tight relationship between the dilemma the user faces (be they provider, patient, or healthcare delivery location such as a hospital) and the specific problem the product solves.

6.2   User-Centered Design: Consumer-Facing Medical Software Requires Finesse

Observation is the best tool for designing medical software applications in healthcare. Too often, the developer and creator of a medical device is situated in an R&D division in the company. A key tenet of product design in the tech marketplace is to go and interact with the customer and, even better, to observe the user in their own environment – see also our discussion on identifying user needs in Chapter 11. There are often numerous stakeholders included in moving an innovation from ground zero to the marketplace. In healthcare these include physicians, patients, nurses, academic institutions, institutional review boards, engineers, information technology (IT) personnel, software developers, human-centered designers, payers, coding experts, and the government (a major payer in many nations). As discussed in this text (see also Section 3.2.3), in healthcare the user and the buyer are two different individuals, and often beautiful design and ease of use, not the feature functionality as often believed, determine the success of a healthcare technology product.

User-centered design takes into account where and how the user will access the information, what the interface will look like, and, in the current healthcare climate, whether it is easy to use both on a laptop, a desktop, and a smartphone (taking into account that images load in a organized fashion and are not distorted, depending on the viewing device, well as secure logins – see Section 3.5 for a brief discussion of cybersecurity). In addition, all of the customer-centric data in retail marketing and the food industry regarding color, font size, and even the way the screen is presented is relevant in healthcare as well. As discussed in the design principles behind a creative design firm like IDEO [118], allowing the designers to create from a clean slate and have the flexibility to alter the interface to best suit the user needs and preferences can help to create a beautiful and easy-to-use product while retaining a tight link between back-end coding and the front-facing interface [118].

A design that makes software easy to use also has safety and security implications, as discussed in Section 13.4.3. A critical component of the risk management process for medical software is to perform a human factors analysis [89] to identify potential hazards that could be caused by deficiencies in design of the user interface.

6.3   Value Proposition: What Do You Bring to the Healthcare Conversation?

What makes us buy the things we buy in a consumer world (e.g. a new car or a new phone) in this day and age is a very important concept to consider when designing for healthcare. While the task is rightfully considered more *serious* or *impactful* in healthcare, given that a surgical tool can change or save a life or cause serious harm (e.g. treatment of a tumor) when used, the principles of paying for a product often remain the same. If you feel you cannot live without the latest phone or exercise gadget for your home, it is because of the benefits and value you ascribe to the product you are thinking about purchasing. For an entrepreneur, it is essential to clearly articulate the value proposition of your product, which is essentially the promise that it will perform a certain set of actions or have features that the user needs. If the user design process has been followed well, this will be clear for the entrepreneur and can lead to a clear set of features and functionality that are beneficial, quite specific, and, most importantly, something that solves a dilemma so the user cannot live without it.

Within the categorization of benefits, there are practical or functional benefits listed as part of the system specifications (see Chapter 12), but also emotional benefits, which can often be just as important. For example, whereas the specific features can determine why a particular telehealth solution provides better-quality images or easier upload of information from the patient, the emotional feeling is what binds the user to the brand, in the same way that brands such as Apple target a feeling of loyalty and a shared belief system. To return to the Pillpack example, they clearly articulated the value proposition of providing an easier system of pill tracking and making the consumers feel secure in their pill-packaging process; these were the functional benefits. At the same time, however, they delighted the customer by easily adjusting the shipment if a prescription was changed, thereby creating in customers an emotional attachment to their brand. Pillpack even had a section of their site with "love letters" from satisfied customers, and this helped validate their model of direct-to-consumer. It often helps a company to have customers create an emotional attachment to their brand and shared values, and Pillpack used this successfully when it was threatened by larger competitors also looking to enter the mail-order pharmacy market. When this occurred, Pillpack alerted their customer base and asked them to lobby their legislators in their local communities and built even greater brand visibility. Monetization of the brand once an early-stage company has built a positive one is a great way to increase valuation of a company and drive investors to the product. In the next sections, we first discuss intellectual property protection and then methods to raise capital. As this particular example shows, building a brand from a grassroots perspective ultimately lead to overcoming large competitors and a very high valuation over time – however, the key is to get capital at the right time as well as to obtain strategic investors.

6.4   Intellectual Property

Intellectual property is the equivalent of signing your name at the bottom of a painting to let the world know, first, that you painted it, second, that you own it, and third, that you have the right to sell it. In that vein, intellectual property, just like a painting, can be forged, stolen and sold, or copied without permission. With medical software design, there is often a perception that it is not possible to protect one's medical software with IP effectively, given that the code can be copied and sold with a different user-interface design with the same functionality. However, that is no longer the case, as medical software is often intimately involved with the functioning of a device, thus allowing for IP protection for the entire product and how it works in healthcare. When designing for healthcare, medical software designers need to be acutely aware that the first target of their competitors will be to copy the software aspect. This is equally true regardless of whether the product lives in the EHR space, the medical device space, or the patient engagement space. There are, however, some key tenets to IP in healthcare, including designing software, that may help you as you navigate this arena. These are discussed below.

6.4.1   File a Patent

In filing a patent, one should include the requirements of why the software is novel, why it is non-obvious, and why it is useful, with the last being the vital application portion of a patent. In the modern era, the Google Patents search engine [101] is a place to start to see what patents exist that are similar either in composition or design, although it is by no means as exhaustive as hiring a patent attorney. Often, legal help is required to provide the broadest protection and to create an IP strategy.

6.4.2   Deciding Whether to be Super-Secret versus Widely Public

There are two schools of thought around IP. One school, followed, for example, by the pharmaceutical industry, says that one should keep all formulations very proprietary and not share anything that might be the "secret sauce." This strategy was evident in the companies competing for the creation of COVID-19 vaccines. In fact, hackers globally have been accused of trying to get at this information. A second school of thought is diametrically opposed and believes that the key to successfully protecting the medical software domain you are creating is making yourself a well-known and well-established player in the space. For example, through provider and patient engagement, publications, and paying customers, frankly, you can make yourself the "market leader," thereby raising the bar for new entrants into the competitive landscape.

6.4.3   Deciding on a Regulatory Strategy

The IP strategy around medical software design is quite tied to the regulatory strategy. If, for example, you are looking to build a niche medical software that enhances the current accuracy of a medical eyeglass assessment system, you can build on the openly available medical software and create something that saves time and is more accurate. Whereas many believe that filing a patent is a necessary first step, as is the utmost secrecy, often the most crucial aspect of a successful venture is the successful execution of your regulatory strategy and getting the first paid customer. Getting FDA clearance, as was discussed earlier in the text (see Section 2.3), is an onerous process and now the FDA has a whole division devoted to digital health and medical software-related healthcare products. The advantage of getting FDA clearance is that it raises the barrier of entry for new companies in the space or even for thieves who stole your painting, made 1,000 forgeries, and are now trying to sell them. That said, though the thieves may not be able to sell the 1,000 forgeries in the United States without FDA clearance of the software, they may be able to export it for value to other countries with less stringent medical regulations. It is critical to understand that FDA clearance is based on the claims around what your invention can do that is useful, so really considering the claims on the IP filing is perhaps the most critical step to take.

One final cautionary piece of advice: It is worth the effort take the time to consult with legal expertise. While IP filings can be very expensive and it is possible to file a patent on your own, even young cash-strapped entrepreneurs will find that many law firms now work for part equity in the business and will be worth their weight in gold should your medical software design be successful. Furthermore, these same law firms often work closely with venture firms and angel investors who are looking for medical software technologies to invest in. This brings us to the next question in the process: How do I raise money for my venture?

6.5   When to Raise Capital: Debt or Equity and What Does It All Mean?

Raising capital, or funding a startup, often involves understanding a myriad of options related to who is giving you money and at what price. This section is intended to provide a very preliminary understanding of these options, as the financial setup can range from the very simple (i.e. self-funded) to the very complex (e.g. a combination of debt and equity).

Often, the earliest start of the venture is funding from friends and family, known as "bootstrapping." Many successful entrepreneurs espouse this philosophy that having a lot of capital means wasting money and being less efficient [145]. Recently, the concept of "family and friends" has really expanded with the advent of crowdfunding, which functions as a way for an entrepreneur to pitch an idea on a platform to complete strangers who invest public funds and often ask for nothing in return. Here again, the *compelling* story can garner enough cash inflow to launch a venture.

Angel investors also present an option for an early-stage venture and can invest in concepts or early design; however, most angel investors require some equity, or a piece of the venture, in return. Equity means ownership in that it represents the investor's percentage of ownership in a venture. It gets more complex in that, for startups, equity means the percentage of the venture's shares sold to an investor for a specific amount of money – and this means that a startup has to decide at this stage what the startup is *worth*, often called its *initial valuation* [220]. An idea at a later stage can be funded by venture capital, usually managed as a fund by professional investors looking for ideas that are more mature. Venture capital can often provide a larger sum of money than all the options mentioned thus far, and in addition they are also *strategic investors*, with knowledge of the area of the startup. In exchange for their shared knowledge and network of contacts to aid the startup, as well as their knowledge of the industry and of how to sell successfully, venture capitalists typically exact more equity [97].

Finally, much more mature ventures with a track record of sales can be positioned to receive money from a private equity firm. Typically, private equity firms invest large sums of money into larger ventures and also expect a much greater return on their investment over a longer period of time. These investments may at times cause turmoil in the venture company, as both venture capitalists and private equity firms may hold so much ownership of the company (or the managing shares) that they have the power to replace the management team.

It is important, therefore, to approach the start of the venture with one's goals clearly in view. If the product being sold and the growth of the company is more important, it may mean that the entrepreneur is willing to give up control and execution of the company as the venture evolves. As mentioned before, these are very financially laden terms but also very emotional topics, as many entrepreneurs become extremely attached to their ideas and their venture, and it is often wise to have a group of trusted advisors while navigating this difficult process.

As one of the reviewers of this book pointed out, it is also worth noting that unlike the rest of the players in this process, the entrepreneur/inventor is the only one for whom this process is a new experience, so they are likely to have a very difficult time selling viable products given the lengthy healthcare cycles. Serial entrepreneurs have the advantage that they learn the "rules of the game" and tend to do better in each iteration, often learning how to navigate directly to the economic buyer.

6.6   Conclusions

An important consideration when analyzing whether a medical software business is viable is to assess whether it solves a large problem the healthcare provider is frustrated by. Once pain points, or dilemmas that cannot be solved with currently available software tools, are identified and validated by interviewing a wide array of providers, then it makes sense to proceed with outlining what software solution may best fit.

This approach allows for testing the market prior to building the tool and allows for iteration and beta versions to be refined prior to coding and regulatory submission. There remains a health appetite for funding healthcare IT ventures and a successful entrepreneur is able to marry the user need tightly with the software solution.

6.7 Summary

- Building a venture in healthcare can be daunting, given the separation between users (providers, patients) and the economic buyers (e.g. the health system), highlighting the importance of the stakeholder mapping process.
- Software solutions face the same legal/IP need and regulatory hurdles as physical medical devices these days, and filing patents as well as understanding the medical claims can be complex and require legal assistance.
- Raising capital may involve seed funding or strategic partnership with investments from a healthcare system – all of which depends on what is most likely to help the entrepreneur build and scale the product.
- The leading indicator of success for funders, once a prototype exists, is actual sales of the product, and the focus on having paying customers at any price cannot be overstated.

RECOMMENDED READING AND RESOURCES

A good introduction to design thinking can be found in this IDEO publication:

IDEO. *Human-Centered Design Toolkit*. IDEO, 2nd ed., 2011.

The following book has an excellent description of the startup process:

D. Kidder and H. Hindi. *The Startup Playbook: Secrets of the Fastest-Growing Startups from Their Founding Entrepreneurs*. Chronicle Books, 2012.

7 Medical Software Applications and Growth Drivers

INTRODUCTION

This chapter presents an overview of current medical software applications and the factors that promise to drive growth in this area. We begin this chapter by defining and discussing some important terms such as digital biomarkers and digital health (Section 7.1). Next, we describe the current technological transformation of healthcare (Section 7.2). In Section 7.3, we present the major challenges faced by the healthcare sector and in particular the triple aim of improving patient experience (better care), improving the health of the population (better quality), and reducing the cost of care (better value). Section 7.4 discusses the promising opportunities that will become available as new technologies are adopted in healthcare, and Sections 7.5–7.10 discuss the drivers of the current wave of healthcare technology growth, including financing, innovations by the Food and Drug Administration (FDA), and increased collaboration across the healthcare continuum. The promises of this technological transformation of healthcare are examined in clinical care, digital medicine, research and development, and remote monitoring. We conclude the chapter with some recommendations for software developers entering this field (Section 7.11)

7.1 Background and Definitions

7.1.1 Categories

Healthcare technology is the broad area of information, analytical, and communications software applications in medicine, research, and other professions to manage illnesses and health risks and promote wellness. It encompasses the use of wearable devices, mobile health, telehealth, health information technology, and telemedicine. The main categories of healthcare technology include [218]:

1. clinical or health information technology (HIT);
2. telemedicine;
3. remote sensing and wearables;
4. data analytics and intelligence;
5. health and wellness behavior modification tools;
6. bioinformatics tools;

7. medical social media;
8. digitized health record platforms and electronic health records (EHRs);
9. patient or physician–patient portals;
10. personal diagnostics;
11. compliance tools;
12. decision support systems;
13. imaging;
14. predictive modeling.

7.1.2 Digital Medicine

The technological transformation of healthcare includes the rapidly growing area of digital medicine, a field concerned with the use of technologies as tools for measurement and intervention in the service of human health. This approach to healthcare is based on high-quality hardware and software products that support research and the practice of medicine broadly, including treatment, recovery, disease prevention, and health promotion for individuals and across populations [52]. For instance, smart pills and ingestible sensors can be prescribed in the treatment of hypertension, diabetes, hypercholesterolemia, heart failure, hepatitis C, mental health, HIV, and tuberculosis, and in organ transplantation. Other examples include disease management software, closed-loop monitoring and drug delivery, and personal health sensors.

7.1.3 Digital Biomarkers

Technology and digital medicine offer better ways to capture patient data through an important area of study called digital biomarkers. Software is used to record data from devices for such information as unique voice characteristics that reflect tremor in Parkinson's disease. Digital biomarkers can include data from electronic patient diaries or clinical outcome surveys. A critical function includes tools that measure adherence and safety, such as chips that can record when medications have been taken, wearable sensors that track falls, or cameras and smart mirrors for monitoring activity in the home. The applications of software as a medical device (SaMD) are increasingly important as diagnostics, including digital biomarkers.

7.1.4 Software–Drug Combination

Software combined with drugs is used to optimize health outcomes. Technological applications are used adjunctively with pharmaceuticals, biologics, devices, or other interventions. The goals of software–drug combinations include the ability to provide patients and healthcare providers with intelligent and accessible software-based tools that deliver high-quality, safe, effective measurements and data-driven interventions. Software is widely used in health research to analyze trends associated with and causes of health and illness by examining data from biological, environmental, and social factors. Observational and interventional clinical trials, as well as health promotion efforts, use a wide variety of technology products.

Healthcare technology that uses software includes interventional products, connected devices, and implantables to prevent, manage, or treat medical disorders or diseases. These can be used independently or together with various medications. Combination products such as continuous glucose monitors (CGMs) are paired with companion apps that have the ability to share diabetic patients' data with their doctors' offices. Other applications could include the recognition of abnormal heart rhythm from an electrocardiogram (ECG) recording from a smartphone or tracing from a smartwatch.

7.1.5 Closed-Loop Systems

Intelligent data recording with software interpretation has led to closed-loop applications with automated responses and less need for human intervention in routine circumstances. For example, programmable insulin pumps must be designed with deep understanding of the biological effects of the injectable drug. This concept is being used in the development of an artificial pancreas, a device that reacts to patients' blood sugar levels and releases appropriately timed doses of medications. These systems can provide continuous monitoring, along with pumps and algorithms that automatically adjust insulin delivery in response to high blood glucose levels (hyperglycemia) and reduce the potential for over-delivery, which may lead to dangerously low blood glucose (hypoglycemia). An example is shown in Figure 7.1.

7.1.6 Data Collection

Novel smart devices can keep highly accurate records of medical interventions. Data pooled from a large population of diabetic patients can be gathered to better understand how small changes in medicine use can affect people so that better guidelines can be designed and implemented. Such evidence-based therapeutic interventions are driven by high-quality software programs.

7.1.7 Digital Wellness

The new concept of digital wellness has been used to describe products that influence consumer well-being, such as exercise trackers or sleep recorders [39]. Some of these products bypass the regulatory approvals process necessary to support medical claims

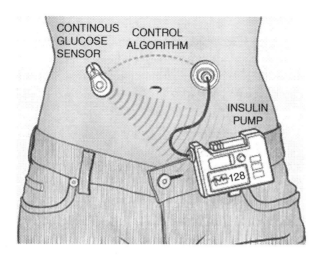

and are intended for consumer use. However, other consumer-grade applications are being tested for use in clinical settings. For example, accelerometers that track motion are becoming more frequently found in clinical trials for medications that may improve patient mobility and activities of daily living (ADL) as secondary or primary endpoints.

7.2   The Technological Transformation of Healthcare

Healthcare is being transformed through the broad adoption of technological applications. Despite the fact that healthcare has lagged behind other industries in the adoption of new technology, recent developments highlight the changes affecting medicine, research, and clinical care. The transformation involves empowering healthcare consumers through the use of clinical or health information technology (see Section 3.3), as well as telecommunications, better and faster computer processing, rapid digitalization of data, and other types of technological innovations. It is being driven by applied technical advances, increased funding, strong regulatory support, and better patient engagement.

7.2.1   Benefits

Software developers should be aware of the causes behind rapid growth of medical technology and the factors influencing the transformation of healthcare through various clinical applications and other solutions. Cautious stakeholders have increasingly recognized that the benefits of adopting new applications outweigh the costs and risks that have been traditionally associated with novel technology. Some of the benefits include promises to improve diagnosis, care delivery, and medical research, to drive efficiencies, and to promote better overall health. Technology is expected to improve the lives of people as active consumers of healthcare. The reason medicine has historically lagged behind other industries in technology adoption has been attributed to the conservative nature of researchers, who cite the high safety bar and expectations set by regulators. Indeed, a reasonable evidence-based benefit-to-risk ratio must be demonstrated to regulatory authorities to receive product approval or clearance in many cases. However, the need for change due to escalating costs in the presence of inferior quality, as well as ubiquitous reliable technology platforms, has helped fuel the transformation.

7.2.2   Technical Advances

The healthcare industry has begun to incorporate various advances in areas such as telemedicine, computing speed, artificial intelligence (AI – see Section 1.3), and use of EHRs (see Section 3.3.1), which have become more accessible and affordable. Better software has played a significant role in driving acceptance as applications continue to more accurately address the needs of practitioners and consumers of healthcare. Greatly increased computer processing power has created new possibilities for analyzing vast healthcare databases.

7.2.3   Improving Healthcare

One of the major drivers of the technological transformation of healthcare is the recognition that there is significant room for improvement. A large body of evidence indicates that the industry continues to face challenges with respect to quality, access, costs, and other issues. In recent decades, costs have escalated to unprecedented levels, errors are common, and quality standards do not match expectations, particularly when examining the amount paid and the level of care that is delivered. There continue to be concerns that the output of the healthcare industry does not meet desirable levels in terms of cost, quality, and access. Among the proposed solutions to improving healthcare are the adoption of novel technologies.

7.2.4   Growth Drivers

Drivers of technological growth in healthcare include investments in:

1. clinical delivery applications at hospitals, medical practices, and individual clinicians;
2. product developers, including pharmaceutical, biotechnology, digital health, medical devices, and software;

3. payers, including insurance plans and governments;
4. research, including academic and industry;
5. investors, including venture capitalists, private equity, and public markets;
6. ability to analyze large aggregated pools of health data; and
7. uptake by patients and healthy individuals.

Key metrics include the increased number of successful health technology and digital health startups that have grown through mergers and initial public offerings (IPOs), and more published studies that demonstrate positive outcomes of health technology applications in disease management.

7.2.5 Solutions

Technological innovation has been heralded as a part of the solution to the healthcare crisis and has been gaining momentum because it is envisioned to [218]:

1. lower the cost of healthcare;
2. reduce inefficiencies in the healthcare system;
3. improve the quality of care;
4. improve access to healthcare; and
5. provide more personalized healthcare for patients.

7.2.6 Professional Groups

As these changes occur, industry leaders and professionals have begun partnering to foster best practices and to support business models. In recent years groups such as the Digital Medicine Society (DiMe [52]) and Digital Therapeutics Alliance (DTA [50]) have been founded to help set guidelines for the development of effective and safe treatments. The Clinical Trial Transformation Initiative (CTTI [40]) works with regulators, industry, and academics to educate and advise stakeholders in best-practice innovations. ASCO CancerLinQ [12] is working in the area of cancer quality improvement. Software developers should understand the applications and drivers of the technological and digital transformation of healthcare. Engineers should also be aware of the key areas affected by changing technologies in order to proceed responsibly and help stakeholders deliver on the promises of these opportunities.

7.3 Healthcare Challenges

7.3.1 High Costs

Digital medicine and healthcare technology products hold great promise to address issues in escalating medical costs, uneven access, problematic diagnosis, and treatment

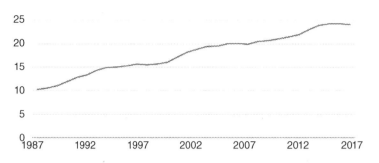

Figure 7.2   Public healthcare expenditures as share of US government spending 1982–2018. Reproduced from
Nunn et al. [191]. Used by permission from the Brookings Institution.

challenges. The problems in healthcare have become national concerns. In 2018, the
United States spent about $3.6 trillion on healthcare, which averaged to $11,000 per
person, or about 18 percent of gross domestic product (GDP) (Figure 7.2). The Centers
for Medicare and Medicaid Services (CMS) projected that within 10 years costs will
climb to $6.2 trillion, or about $18,000 per person, and will represent about 20 percent
of GDP. Healthcare also became the largest employment sector in the United States
in 2018 [253]. Despite extraordinarily high costs and utilization of expensive services,
there is also disappointing quality as measured by factors such as disease prevention,
infant mortality, and equitable access to care.

7.3.2   Quality Concerns

Unfortunately, increased spending in healthcare is not necessarily correlated with better
patient outcomes [117]. Despite leading the world in costs, the United States ranks
26th in the world for life expectancy and ranks poorly on other indicators of quality.
Estimates are that per capita spending on healthcare is 50–200 percent greater than in
other economically developed countries, raising questions about the amount of waste in
the system [25], although there have been attempts to control escalating costs through
reform, such as the 2010 Affordable Care Act (ACA – the deployment of the online
marketplace for the ACA is discussed in Chapter 19). The ACA provided coverage
expansion and organic focus on the "triple aim," defined as better care, better health,
and better value, but much still needs to be done [265].

In Europe, the Organisation for Economic Co-operation and Development (OECD)
has focused on quality of care. It reports [194] large variation in care outcomes both
within and across countries. For example, avoidable hospital admissions for chronic
conditions such as asthma and chronic obstructive pulmonary disease (COPD) can vary
by a factor of nearly 10 between countries.

At the international level, quality is receiving increasing attention through the
Sustainable Development Goals (SDGs). The World Health Organization (WHO)
reports facilitate the global understanding of quality as part of universal health coverage

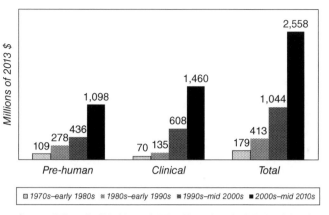

Sources: 1970s–early 1980s, Hansen (1979); 1980s–early 1990s, DiMasi et al. (1991); 1990s–mid 2000s, DiMasi et al. (2003); 2000s–mid 2010s, Current Study

Figure 7.3 Drug development costs. The cost of developing a new drug has skyrocketed since the 1970s. Figure reprinted from Joseph A. DiMasi et al. [51]. Used by permission from Elsevier.

aspirations and include the SDG imperative to "achieve universal health coverage, including financial risk protection, access to quality essential health-care services and access to safe, effective, quality and affordable essential medicines and vaccines for all" [272].

7.3.3 Drug Development

Drug development is extremely expensive, with current estimates ranging between about $1 billion and $2.6 billion to bring a drug through discovery, testing, and approval [16] (Figure 7.3, see also Mulin [183]). The median estimates for different therapeutic areas range from about $760 million for the nervous system to $2.8 billion for cancer and immunomodulating drugs [112]. Other studies indicated that the median research and development investment to bring a new drug to market was estimated at $985 million to $1.3 billion [251]. Surprisingly, the rate of return on pharmaceutical company R&D investment dollars is just 3.2 percent annually due to the high failure rate of drug development [45]. Experts have expressed concern that high development costs are driving up prices for drugs and threatening the financial sustainability of providing treatments in specialized areas such as cancer [271].

7.3.4 Cautions

Regulatory concerns are often cited as the reason decision-makers in the healthcare and pharmaceutical industries have been cautious to adopt new technologies. Additionally, incumbents in healthcare and biopharmaceuticals tend not to have the organizational, cultural, and technical capabilities to embrace new technologies. In order to gain approval, products must be shown to have therapeutic benefit with a minimum of adverse effects. However, the need to find better methods for basic science research and drug discovery processes has led to increased use of sophisticated analytics, genomics, and biomarker research. In the commercialization area, the use of digital marketing and software-based data analytics has become commonplace.

7.4   Promises of Healthcare Technology

As new technologies are adopted in healthcare, industry experts say extremely promising opportunities will become available. Technology has the possibility to achieve the remarkable combination of "driving down costs and empowering consumers to take charge of their health," according to Donald Jones, the Chief Digital Officer of the Scripps Research Translational Institute (SRTI) [236].

7.4.1   Benefits

The objectives of digital health technology products and services have been described as [218]:

1. improve the quality of outcomes of care and services;
2. improve population health;
3. improve the patient experience;
4. improve the physician and other non-physician provider experience; and
5. address health disparities.

7.4.2   Better Data

Many industry professionals and academics believe that part of the solution to industry challenges could be in the application of better data analytics in healthcare. According to Bernard Tyson, the former CEO of Kaiser Permanente, the basis of that organization's success in delivering quality at a lower cost was sophisticated pooling and analytics of patient data and use of medical informatics [232]. The use of EHRs is approaching universality in the United States and other parts of the world. Many health systems use the Internet and mobile applications to better connect patients with their healthcare systems and providers.

7.4.3   Mobility

Medical futurists predict that digital technologies will allow for the seamless flow of data, ranging from healthcare records and systems to patients interacting with care teams at any time and from anywhere. Personal data records will become more transportable and accessible. Telemedicine is expected to become more routine mobile care, similar to the ease of using a cell phone or iPad. Such telemedicine applications have experienced a rapid rise during times of social distancing, quarantines, and isolation, such as during the COVID-19 pandemic of 2020.

7.4.4   Drug Development

The promises of digital technologies in the drug development process are multiple. Making new drugs is a highly complicated, multi-stage process. At its core, drug development involves the creation of new data and the analysis of that data in order to determine whether novel medications are safe and effective. Digital and mobile applications can bring together physicians and potential subjects for clinical trials more easily. Mobile platforms can be used to collect data and then efficiently store it in centralized databases for later analysis. Real-world evidence (RWE) includes all of the positive and negative information about medication use in the "real world" after clinical trials are complete and products have been approved. It can help facilitate drug development as a better understanding of baseline and comparators. Digital technologies can help make drug development more efficient by increasing the speed of enrollment, distributing information, getting more patients eligible to participate, and collecting relevant data quickly and accurately. The cumulative effect is expected to be an overall reduction in development time as well as potentially overall cost reduction in drug development.

7.5   Technology Growth Drivers

7.5.1   Digital Future

The "digital future" of healthcare has been heralded by multiple authors. Perhaps one of the best regarded is physician futurist Eric Topol, MD, founder and director of the Scripps Research Translational Institute in La Jolla, California. Topol's independent report [256] outlined how the hospital workforce should prepare for a technological transformation that will require approximately 90 percent of all jobs in healthcare to have deep digital skills. Within the next decade, staff will need to be able to navigate a data-rich healthcare environment that requires digital and genomic literacy. The top 10 digital health technologies to impact the hospital workforce in the next decade, as outlined by Dr. Topol, are [256]:

1. telemedicine;
2. smartphone apps;
3. sensors and wearables for diagnostics and remote monitoring;
4. reading the genome;

5. speech recognition and natural-language processing;
6. virtual and augmented reality;
7. automated image interpretation using AI;
8. interventional and rehabilitative robotics;
9. predictive analytics using AI; and
10. writing the genome.

7.5.2   Patient Empowerment

Global pharmaceutical companies, payers, and insurance companies have recognized the business benefits of patient empowerment. Patient-oriented applications help healthcare consumers become decision-makers for their own health by providing tailored recommendations and personalized medical information. Thus, expanded engagement through apps that help patients become better medical decision-makers drives healthcare technology adoption.

7.5.3   Disease Management

A recent report by the World Health Organization (WHO) notes that the global burden of chronic disease is growing. This burden is a major challenge for healthcare systems around the world, which have largely developed to deal with acute episodic care rather than long-term conditions [273]. Chronic disease management systems (CDMS) are software applications for healthcare data sources to monitor and process patient information in order to help manage chronic and subacute disease. They are used by hospitals, clinics, and other healthcare practices to engage, educate, and treat patients. These systems integrate patient records, charts, medications, test results, and medical decisions to provide summary reports, which include updates on treatment plans and recommendations to improve the quality and thoroughness of medical care. Benefits include optimization of self-care, reduction of healthcare expenditures, appointment scheduling, systematic reminders for medications, medical procedures, blood draws, and education for healthier behavior regarding food, exercise, and other personal choices.

Around 50 million people in Europe suffer from multiple chronic diseases, and more than half a million individuals of working age die of them every year, representing an annual cost of some € 115 billion [195]. Patients are becoming more enthusiastic about using digital applications and the Internet for medical information regarding chronic diseases. A recent publication showed that more than half of patients surveyed reported a preference for receiving support on mobile platforms. Most cardiac patients stated that they would be interested in receiving additional health support using email, websites, and online videos [26].

Several notable companies have made progress in disease management applications. Livongo Health was started in 2014 with an initial focus on diabetes management.

The company broadened its platform to include diabetes prevention, behavioral health, and hypertension through developing internal technologies and acquisition of connected device, data, and clinical capabilities. Recently Livongo announced merger proceedings with Teladoc Health, a multinational telemedicine and virtual healthcare company that operates in 130 countries and serves around 27 million members [249]. The combined Livongo–Teladoc entity is projected to have a market value of over $18 billion [157]. See Section 7.5.5 for more.

Pear Therapeutics received regulatory approval for prescription digital therapeutics in September 2017 [204]. This was the first FDA-cleared software application with claims to improve clinical outcomes for a disease. The RESET program demonstrated improved abstinence and treatment retention in a randomized controlled clinical study in patients with substance use disorder (SUD). The digital therapeutic developer Voluntis received marketing authorization from the FDA in 2019 for the first SaMD (see Section 2.2.3) cleared for use against different types of cancer [262]. Voluntis is a care companion for cancer patients' journeys based on mobile and web technologies. It uses embedded clinical algorithms, patient symptom self-management, and remote patient monitoring partnered with care team supervision.

The increased interest in consumer engagement and availability of smartphones has led to many healthcare startups developing apps to improve patient care. There were over 318,000 health apps available in 2017, with more than 200 apps being added each day, including fitness, diagnostics, and symptom management. [121]. Demonstration of app efficacy is being continuously confirmed in a growing body of evidence, including over 570 published studies.

7.5.4 Venture Capital

Fueled by revenue growth expectations, the global market for telehealth, apps, health analytics, mHealth, and digital health systems has been projected to reach over $500 billion within a few years [4]. Private sources such as venture capital (see Sections 6.1 and 6.5) have been driving investment in health technology. Venture capital funding was about $8 billion in 2018 and about $7 billion the following year, and increased to over $5 billion for the first half of 2020 [32].

7.5.5 Public Markets

Digital health technology companies have been financed through initial public offerings (IPOs – see Section 6.5) to stock markets. In 2018 and 2019, IPOs resulted in a combined market value of $17 billion. Notables included Livongo; Change Healthcare (CHNG), which sells software and tech-enabled services; Phreesia (PHR), a maker of patient check-in software; and Health Catalyst (HCAT), which develops population health management solutions. Other public companies, such as cloud-computing Veeva

(VEEV) have seen their stock soar. As noted, Teladoc and Livongo initiated a merger valued at over $18 billion for a company that was just 12 years old [157].

7.5.6 Partnerships

Additional resources have been flowing from large companies that fund innovation and alliances. Novartis, one of the world's largest pharmaceutical companies, has been embracing technology by creating the new role of chief digital officer (CDO). Company CEO Vas Narasimhan has stated that his goal is to be the "leading medicines company, powered by data and digital." The pharmaceutical company established a digital innovation lab called the Novartis Biome with the specific intention of empowering health tech companies and fostering partnerships [190]. The first Novartis Biome was opened in 2019 in San Francisco, and another in Paris, with more planned in the coming years. At industry meetings such as HLTH and Digital Therapeutics, held on the Harvard Medical School campus, major companies set expectations for partnerships that would take advantage of patient consumerism trends [44].

7.6 Healthcare Data

7.6.1 Data Markets

The interpretation of trends from aggregated and then anonymized medical records, prescriptions, and payer data are highly valued by academic researchers, manufacturers, and drug makers. The market for consumer health data is estimated to be valued at between $50 billion and $100 billion, and growing 30 percent annually, according to industry experts and CEOs. Pharmaceutical companies can spend tens of millions of dollars each year acquiring health data for commercial purposes [116]. This has led to high sales such as the analytics company Optum Insights, which had revenue totaling approximately $10 billion [275].

7.6.2 Public Health

The ability to study trends has long been a mainstay in understanding diseases. Public health and policy experts have found that examining healthcare data can lead to important insights about how different populations receive care and about how much they or their payers are charged. Over time research into health trends has led to public policy implications such as legislation, sanitation, and funding. For example, studies using Medicare data have found racial disparities in access, treatment, and readmissions following surgery [259]. Further work led to the observation of huge price variation in private insurance for the same service at the same hospital.

The Pharmaceutical Group in the European Union (PGEU) has recommended that EHRs should be linked with electronic systems that would allow pharmacists to securely access and contribute to patient information and provide medication information. The goal would be to promote patient safety, improve quality of care, and avoid errors and duplication of treatments. PGEU notes that "pharmacy apps" feature integrated reminders and alerts that help patients improve self-care and ensure medication adherence [207].

7.6.3   Medical Informatics

Medical informatics is the analysis of healthcare data using computers, health information, and cognitive sciences to medical insights. The field is based on pooling and analysis of health records, conditions, treatments, and outcomes, combined with genetic data, information on local environmental conditions, and exercise and lifestyle habits. Software-based analytic programs are needed to enable the effective application of medical informatics using data science.

7.6.4   Health Gains

It is believed that substantial gains could be achieved for the healthcare industry through more widespread pooling and analysis of medical data. Examples of the benefits of medical analytics include better understanding of treatments and drug dosages, such as which combinations work best, how different medical treatments interact, if genetic markers are predictive of success, and which life choices improve health [209]. Additionally, price drivers could be better understood, leading to more efficient use of resources, which could reduce overall healthcare costs.

7.6.5   Data Growth

Medical knowledge has been growing so quickly that it is now expanding exponentially. It has been estimated that it took about 50 years for all of the world's medical knowledge to double during the early twentieth century. This doubling time accelerated to 7 years in 1980, 3.5 years in 2010, and is now projected to be only 73 days [46]. If this pace continues, all the medical data that exists today will represent only 10 percent of the data pool of the next decade. Therefore, better and faster ways to analyze larger healthcare databases are needed.

7.6.6   Data Privacy

There are significant benefits to collecting and analyzing medical data, but these need to be balanced by the need to ensure personal privacy and maintain data health security. Corporations routinely collect and de-identify vast pools of prescription information, insurance payments, and medical records and resell aggregated data for commercial purposes. The ethics of personal data rights versus the public benefit of aggregating data for general welfare is a complex issue. Tension between the public benefits of aggregated, shared data and the right to privacy continues to be studied by academics and policy experts. See also the discussion on HIPAA in Section 3.4.1 and GDPR in Section 3.4.3.

7.7   Technology in Clinical Trials

7.7.1   Electronic Health Records

Technology advances should improve clinical research in multiple ways. Patient screening will be facilitated by analyzing data in EHRs (see Section 3.3.1). An advantage would be to expand the pool of eligible participants and increase the diversity of a trial population. Medication adherence can be better tracked and improved using smart pill packages for investigational drugs. Wearable devices are able to record data continuously and better capture safety events. For example, smartwatches with cardiac monitoring capability are able to record infrequent heart problems, such as rare arrhythmias, that might otherwise be missed.

7.7.2   Decentralized Trials

Interest in a new type of mobile, decentralized clinical trial has been mounting. These trials are conducted using mobile technology, with patients participating from home with few or no hospital visits. The clinical development process is extremely slow and expensive, with costs reaching over $2 billion [51]. Industry leaders have been searching for ways to drive efficiencies. The virtual approach was initially used for passively collecting patient information. The next step was to use mobile diagnostic capabilities. The landmark Apple Heart Study recruited about 400,000 participants in record time and showed how wearable technology can help detect atrial fibrillation [238]. Johnson & Johnson (JNJ) partnered with Apple for its Heartline Study on stroke prevention and plans to recruit over 150,000 participants [139]. The field progressed to drug approval trials in a fully remote and virtual format with the announcement of a trial called CHIEF-HF (Figure 7.4) that has the goal of evaluating diabetes drug Invokana in patients with heart failure [137].

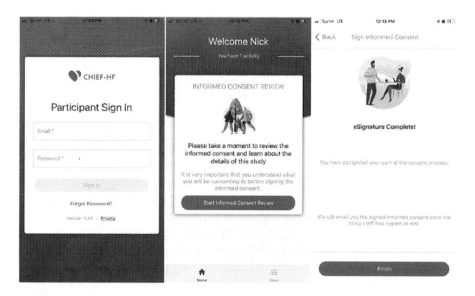

Figure 7.4   The CHIEF-HF mobile application. This is a mobile health platform (mHP) solution to tracking participant informed consent, enrollment, and study activities.

7.8   Regulatory Drivers

The goal of digital therapeutics is to deliver evidence-based therapeutic interventions through high-quality software programs to prevent, manage, or treat medical disorders or diseases. They are reviewed and cleared by regulatory bodies as required to support product claims regarding risk, efficacy, and intended use.

7.8.1   US FDA Modernizations

The FDA (see Chapter 2) has demonstrated dedication to promoting innovation in order to accelerate product development with a renewed focus on data and technology. The 21st Century Cures Act of 2016 was a key catalyst in the use of real-world data and patient-focused drug development. The strategy was expanded with the FDA Technology Modernization Action Plan of 2019 [67], which outlines how the agency will work to ensure data security, develop a series of living examples and use cases, and learn how to communicate and collaborate with external technology innovators to build better products (see Section 2.6). Digital medicines are software-driven connected technologies that include therapeutics, as well as measurement, diagnostic, and monitoring tools.

The agency published a Digital Health Action Plan in 2017 that "re-imagines FDA's approach for ... timely access to high-quality, safe and effective digital health products" [90]. In 2018, the FDA approved the world's first software as a therapeutic with the new designation software as a medical device (see Section 2.2.3). This was a software application for non-opioid drug addiction that was developed by Pear Therapeutics [204]. The event was the culmination of years of work, as well as collaborations between innovative companies and regulators in a unique program called the Digital Health Software Precertification (Pre-Cert) Program [85] (see also the discussion in Section 4.1). This voluntary effort created a more tailored approach to new technology by looking first at the technology developer rather than the product in order to establish the most appropriate criteria for supporting a firm-based program.

The FDA's dedication to technological innovation has culminated with the launch of the Digital Health Center of Excellence within the Center for Devices and Radiological Health [68]. Its goal is to further the agency's dedication to the advancement of digital health technology, including mobile health devices, SaMD, wearables used as medical devices, and technologies used to study medical products. It will help continue the effort to modernize digital health policies and regulatory approaches and provide efficient access to highly specialized expertise, knowledge, and tools to accelerate access to safe and effective digital health technology.

7.8.2 European Medicines Agency

Leadership at the European Medicines Agency (EMA) has noted that digital technologies are becoming part of the conduct of clinical trials. In order to provide recommendations and avoid confusion, the organization has been issuing guidance documents regarding digital methodologies to support product approval. In June 2020, the EMA provided clarification and defined a digital endpoint as a precisely defined variable intended to reflect an outcome of interest that is statistically analyzed to address a particular research question, and may include digital biomarkers, or electronic clinical outcome assessment (eCOA) [63].

The EMA and Heads of Medicines Agencies (HMA) published their Joint Big Data Taskforce's summary in February 2019 [108]. The report laid out recommendations for "rapidly accumulating" extremely large data sets that can be used to computationally analyze and "reveal patterns, trends, and associations" in support of the evaluation and supervision of medicines by regulators. The report prioritized clinical trials and imaging, RWE, adverse drug reports, and social media and mobile health.

7.9 Industry Organizations

Professional associations and quality standards are necessary to help promote adoption of novel fields such as digital therapeutics. This rapidly emerging industry is bringing together regulators, engineers, practitioners, and companies to agree on definitions that enable innovation, according to product development and regulatory specialists.

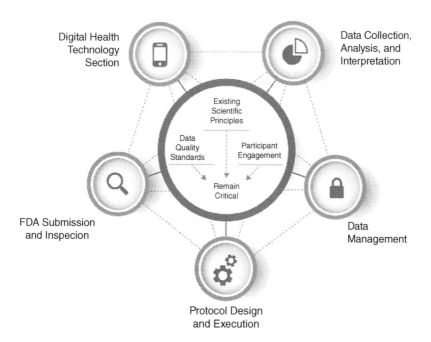

Figure 7.5 Clinical Trial Transformation Initiative (CTTI). The goal of the CTTI [40] is to advance the use of digital health technologies for data capture and improved clinical trials. Image from ctti-clinicaltrials.org.

7.9.1 Education and Policy

The FDA has been working with multiple policy organizations, among them the nonprofit Clinical Trial Transformation Initiative (CTTI; Figure 7.5), which is funded by the FDA and the Duke Margolis Foundation [40]. This organization brings together leaders from industry and academia, as well as regulators, to provide policy advice and recommendations. For example, the CTTI's summary recommendations concerning mobile trial technology include device selection, data collection, and management and regulatory requirements [42]. The group also provides novel drug development strategies, such as for Decentralized Clinical Trials [41]. Another major initiative has been to provide recommendations for the use of RWE in the clinical trial process [43].

7.9.2 Professional Groups

Professional groups and industry-wide collaborations have emerged to drive the growth of healthcare innovation. The development of high-quality, evidence-based products and services is now being supported by the Digital Medicine Society (DiMe [52]), as well as the DTA [50]. Both organizations share the long-term vision of helping patients and clinicians meaningfully use and benefit from these products and digital solutions.

The DTA industry association and DiMe professional society were founded to develop industry foundations through product category definitions, standards of quality, safety, and evidence generation, in addition to frameworks for scalability and sustainability. DiMe supports and educates professionals, whereas DTA helps the industry at the organizational level. Together, these two groups represent the internal drive within a rapidly evolving industry to ensure that high-quality, evidence-based products are delivered to end users.

New technologies must be effective, reliable, and high quality in order to be adopted by practitioners. The FDA is charged with clearing products based on safety and efficacy data. Helping the industry to generate evidence-based and patient-centered standards is being facilitated by groups such as the CTTI, DiMe, and DTA. With appropriate education and standards, digital health has the potential to empower patients through increased access to high-quality care, better clinical results, and better economic outcomes. Without appropriate validation and security standards, digital health technologies could become ineffective or even pose inadvertent safety risks.

7.10   Remote Monitoring

7.10.1   Digital Monitoring

The variety of healthcare technology monitoring capabilities is growing. Digital measurements are relatively well-established in routine clinical care and are moving toward smaller, more patient-friendly applications. The development of electroencephalograms (EEGs) and electrocardiograms (ECGs) are examples of successful biometric sensors from the last century used to analyze signals from the brain and heart. Now, the use of ambulatory ECG monitoring to detect arrhythmias has begun to shift from bulky Holter monitors that require multiple wires toward more convenient smartwatches or other portable connected devices. Remotely monitoring patients with implanted heart devices allows doctors to better follow their cardiac patients and to detect abnormal heart rhythms.

7.10.2   Patient Safety

Real-time safety monitoring, including fall detection systems, is designed to protect elderly and frail populations. Such systems include wearable sensors, cameras, motion sensors, and microphones. Post-operative recovery patients are now using app-enabled wearable sensors to improve physical rehabilitation. Range of motion and step count can be tracked remotely to record patient progress.

In order to improve treatment adherence, a digital ingestible sensor embedded in a medication capsule has been developed. The innovative system to track when pills are taken was co-developed by Otsuka Pharmaceuticals and Proteus Digital Health and was approved in 2017 for use with a behavioral health medicine [197]. Other adherence solutions include sensors integrated into the packaging of medicines to record when drugs are administered.

7.11 Recommendations

Software developers should be optimistic about the magnitude of technology and digital medicine adoption in the healthcare industry. There are many exciting positive developments in this industry that touches so many people, and there continues to be room for improvement. Executives and decision-makers should move quickly and strategically, while remaining conscious of their safety responsibilities. The following considerations are guidelines for executives and software developers.

1. **Patient safety first:** Safety considerations should be foremost and the highest standards need to be applied (see Chapter 4). Additionally, any technology that connects to the patient should be carefully thought through in order to make sure that users have clear understanding and are able to easily use the technology. Often, existing platforms such as mobile phones may be easily adapted for such purposes.
2. **Invest in solutions, not technologies:** Decision-makers should ensure they evaluate the overall solution and business case, not just become enamored with the latest flashy technology. Use cases that provide tangible solutions for patient and healthcare consumers should be prioritized.
3. **Consider pilot programs:** Most products and development programs should have a test case in order to demonstrate proof of concept before moving into full-scale programs. Small market research studies or pilot programs provide valuable feedback.
4. **Find early wins:** The first work for any organization should be to keep it simple in order to demonstrate that a new technology can be effective. The best way to proceed is in a focused manner that shows how a technical solution can achieve results before moving onto larger and more complicated or expensive programs.
5. **Therapy selection:** Some technologies are better suited for specific therapeutic areas. Technology pioneers have the ability to establish success criteria tailored to specific conditions. For example, online digital interventions initiated in the field of behavioral health allow patients and caregivers to more easily communicate remotely.
6. **Focus:** Define the market and the customer base. The flow of money in the US healthcare system can be convoluted. Innovators must understand the complexities to understand patient needs and clearly define market opportunities.

7.12 Conclusions

We are fortunate to be witnessing an exciting transformation as the healthcare industry embraces technology. The pharmaceutical industry is investing heavily and becoming a major player in the digital movement. These changes bring expectations of significant improvements in the quality of healthcare delivery and patient engagement and well-being. Standalone applications and software–drug combinations are gaining wider acceptance. Sophisticated data analytics to capture insights from EHRs, personal health records, and patient portals to support clinical decision-making are becoming more common. Digital measurements will help inform better clinical care as well as bring more efficiencies to clinical trials and drug development. The software developer has the unique ability to fundamentally contribute to these advances.

7.13   Summary

- Advances in medical technology are expected to improve quality and reduce costs of healthcare and accelerate product development in multiple ways. For example, recent progress in telehealth, digital health solutions, and computerized innovations have been helping refine diagnostics, treatments, provide point-of-care testing, and support more sophisticated clinical research.
- Additionally, there are increased opportunities to capitalize on the improved delivery of healthcare, particularly in remote settings. A positive future vision is being driven by technology and includes improvements in public health with better diagnoses, patient empowerment, and more equitable access to care across populations.
- Software is central to all of these types of medical applications. Multiple applications of software, drug–software, and therapy–software combinations exist and will continue to evolve.
- There are a variety of expanding uses for software across healthcare. Important growing categories include digital medicine, telehealth, SaMD, disease management software, and healthcare data analytics using AI and ML.
- Academic organizations and regulators are creating new structures including professional groups and societies. These entities partner with businesses to develop safety and quality standards that are necessary to promote reliable product development and ethical growth in emerging areas such as digital medicine.
- The broad incorporation of technology into healthcare is driven by user and patient needs, and fueled by technical advances and financial investments. These growing changes are expected to continue to reduce overall costs, improve health and outcomes, and provide more equitable access to care.

RECOMMENDED READING AND RESOURCES

For an introduction to the topic of digital health, see:

Y. Ronquillo, A. Meyers, and S.J. Korvek. Digital health. *StatPearls (Internet)*, July 4, 2020.

For an overview of creating successful new products and services, see:

R. Herzlinger. *Innovating in Healthcare: Creating Breakthrough Services, Products, and Business Models*. Wiley, 2021.

For a description of the exploding costs of healthcare, see:

D. Thompson. Health care just became the U.S.'s largest employer. *The Atlantic*, January 9, 2018. www.theatlantic.com/business/archive/2018/01/health-care-america-jobs/550079.

A description of how digital technologies are poised to transform medicine can be found in:

E. Topol. *Deep Medicine: How Artificial Intelligence Can Make Healthcare Human Again*. Basic Books, 2019.

E. Topol. The Topol Review: an independent report on behalf of the Secretary of State for Health and Social Care. NHS, Health Education England, February 2019. https://topol.hee.nhs.uk/wp-content/uploads/HEE-Topol-Review-2019.pdf.

Medical regulators are also actively thinking about this type of transformation. For good examples of this type of their planning, see:

FDA, Center for Devices and Radiological Health. Digital health innovation plan, 2020. www.fda.gov/media/106331/download.

European Medicines Agency. Questions and answers: qualification of digital technology-based methodologies to support approval of medicinal products, June 1, 2020. www.ema.europa.eu/en/documents/other/questions-answers-qualification-digital-technology-based-methodologies-support-approval-medicinal_en.pdf

For good articles from professional digital medicine and digital therapeutics organizations, see the web pages of the following two organizations:

Digital Medicine Society (DiMe) Publications. www.dimesociety.org/research/dime-publications/

Digital Therapeutic Alliance (DTx) Foundational Documents. https://dtxalliance.org/understanding-dtx/

Part II

Scientific and Technical Background

8 Mathematical Background

INTRODUCTION

Much of the work involved in the development of medical software (and in particular the process of software validation) depends critically on an understanding of topics such as probability theory, statistics, and increasingly machine learning. The goal of this chapter is to provide students with some theoretical grounding in this general area. After a brief high-level introduction to decision theory (Section 8.1), we discuss each topic in turn, beginning with an introduction to probability (Section 8.2) and statistics (Section 8.3), including an extended discussion of the topic of signal detection, which is particularly important in medical applications. We then present an overview of machine learning and some of the issues involved in the use of these techniques (Section 8.4). Next, we introduce the concept of statistical significance (Section 8.5) and conclude the chapter with a discussion of the importance of randomization (Section 8.6) in clinical studies.

8.1 Data-Based Decision-Making

In many aspects of system/software-design, one needs to make decisions. These decisions could be made automatically, using an algorithm, or manually (e.g. choosing some particular strategy). The general process is illustrated in the schematic shown in Figure 8.1. We start with the real world on the right. If, for example, we are designing software to detect brain tumors from magnetic resonance images (MRI), our real world is a collection of patients suspected of having brain cancer. We then use (primarily) software tools (whether automated or interactive) to identify the presence or absence of a tumor (and any other characteristics, such as volume and shape) from images of these patients. This process, captured in the middle box, results in imperfect measurements (due to potential limitations in both the images and the software) and potentially some predictions as to the likelihood of this patient surviving for longer than a certain period of time. The right-hand box defines the area in which our scientific knowledge comes into play. Once this is completed, we take the results of our scientific investigation (ideally together with the uncertainties in our estimates/models) and use them to make a decision about how the patient should be treated.

Figure 8.1   Data-based decision making flowchart. First we create data by a combination of measurements and models. Next, we use this information to arrive at a decision. In the middle box, one can find all the usual tools from statistical measurements (e.g. feature extraction, statistical estimation, event detection, classification), while the rightmost box is mostly concerned with ethics and values.

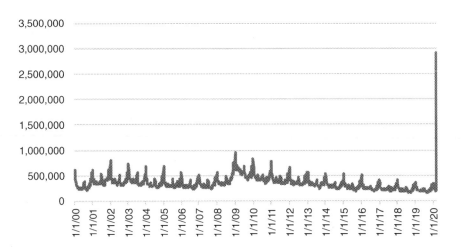

Figure 8.2   The data will not tell you everything! This plot shows the weekly jobless claims in the United States for the period 2000–2020, which included the economic collapse of 2007–2008 [181]. Nothing in this data could be used to predict what actually happened at the start of the COVID-19 pandemic. Data source: US Department of Labor.

Recently, comments regarding the use of "the Science" or "the Data" to make a decision have become very common (and some for good reasons). One has to be clear, however, that while our measurements and scientific knowledge should provide the inputs for the decision, the actual decision depends on non-scientific issues such as morals/values/ethics. For example, an older patient may have a higher probability of survival if they have surgery, but they may feel that the more invasive procedure is unnecessary at this stage of their lives. This may result in them being more interested in palliative care instead.

A final point that is worth addressing here is the "Silicon Valley attitude" that the "data will tell you everything." This has become widespread as the availability of big data/large data sets has moved many fields from model-based analysis to pure data-driven analyses. This is greatly facilitated by emerging deep learning techniques for data analysis. Such analyses have been very successful recently, leading many to believe the old world of probability, statistics, and modeling is history. Figure 8.2

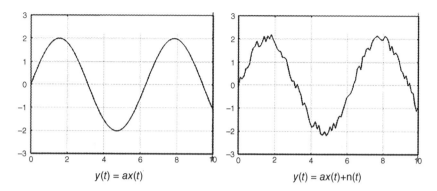

$$y(t) = ax(t) \qquad\qquad y(t) = ax(t)+n(t)$$

Figure 8.3 Deterministic vs. random system. Shown here is a comparison between a deterministic system (left) and a random system (right), where $x(t)$ is the sine wave function. Note that, for the deterministic system, if one knows a (set to 2.0 in this case) and $x(t)$, one computes $y(t)$ perfectly. This is not the case for the random system on the right, where we can estimate the value of $y(t)$ subject to some uncertainty.

provides a counterexample. Nothing in the last 20 years would have led one (or one's algorithm) to believe/predict the explosion of unemployment claims seen in March 2020. This should be a cautionary tale: sometimes the unpredictable/seemingly impossible happens.[1]

8.2   A Brief Introduction to Probability Theory

In this section we provide a very cursory, "hand-wavy," introduction to probability theory. This introduction is intentionally less rigorous than those one finds in standard textbooks on the subject (e.g. the classic textbook by Papoulis [200]). For those readers who have had rigorous training in this subject, we ask for forgiveness as we try to provide in a few pages what is often multiple semesters' worth of material.

Two Examples Generally, we use probabilistic modeling to capture and manage uncertainty. We use this to understand/model processes that are not (or at least do not appear to be) fully deterministic. As a first example, let us consider a sample deterministic system that is governed by the following equation (this is also plotted in Figure 8.3 – left):

$$y(t) = ax(t),$$

where t is time, $x(t)$ is the input to some system (e.g. an audio amplifier), $y(t)$ is the output, and a is the parameter of the system (e.g. the volume control). If this model captures our system perfectly, then knowledge of $x(t)$ and a allows us to perfectly compute $y(t)$.

However, in most actual situations we have uncertainty and cannot perfectly compute the output $y(t)$. Our model, for example, may not be a perfect representation of the system. This might be because there are truly random effects (e.g. random noise in the data) or because our system is too complicated to model precisely and a simplified model is good enough for our needs. In both of these cases, we may end up with a situation, shown in Figure 8.3 (right), where:

$$y(t) = ax(t) + n(t),$$

where $n(t)$ is a random process (e.g. noise) and captures either the true randomness of the system or the imperfection of our model. In this scenario we can only estimate $y(t)$ with a certain level of confidence that depends on the properties of the noise/unmodeled variation $n(t)$. Creating such a model requires the use of probabilities and tools from probability and statistical theory.

As a second example, let us consider the case of a fair coin.[2] This results in:

- heads 50 percent of the time; and
- tails 50 percent of the time.

Theoretically, this is a deterministic process. If one could model the mechanics of the coin-flipping process, measure the exerted forces, account for air flow (and potentially temperature and humidity), the height from which the coin is flipped, the properties of the floor, and many many, other variables, one could model the process exactly and be able to compute the outcome of the process. In practice, we cannot measure all of these parameters and, therefore, it may be that the best way to characterize this process is probabilistically.

What Can Probabilistic Modeling be Used For? We can use probabilistic modeling in situations such as the following (this is not an exhaustive list):

- situations that we do not understand/cannot completely measure – this is the coin-flip example;
- situations that we do not control, such as "what is an adversary likely to do?";
- situations that we could model completely but the level of effort required is simply not worth it for our needs. Here we can live with a certain amount of uncertainty provided that we know it is small enough.

More technically, in probability theory we have tools that allow us to handle situations such as the following:

1. How likely is it that event X will occur? For example, how likely is it that it will rain tomorrow?
2. How likely is it that event Y will occur, given that X has occurred? For example, how likely is it that it will rain tomorrow, given that it has rained today?

In the medical world, the typical questions might look like:

1. Is the difference in life expectancy between cancer patients that were given an experimental drug and patients that were given a placebo[3] real or just random luck?
2. How likely is it, if somebody tests positive for a particular condition, that they actually have the condition as opposed to it being a false positive?

Practically, probabilities are estimated from multiple observations. Given a number of observations (measurements), we can create (in a variety of ways) a probabilistic model and then use this model for predicting future outcomes.

8.2.1 Probability Density Functions

Univariate Probability Density Functions Probably the single most important concept in probability theory is the probability density function (pdf). This is a function that captures how likely the occurrence of an event is. This function may be given to us, or we may have to estimate it from actual measurements (see Section 8.3.3). For example, consider the case in which we are interested in the problem of predicting the outcome of a coin flip. Here, if the coin is fair, we can define the pdf as follows:

1. The domain of the function or the list of potential events is simply *heads* or *tails*.
2. We define the outcome as the variable x.
3. We can then write the pdf as $p(x = outcome)$, where *outcome* is either heads or tails.
4. So the pdf in our case is: $p(x = heads) = 0.5$ and $p(x = tail) = 0.5$. This is known as a binomial pdf.

This is an example of a discrete pdf. In this case, the event space consists of two distinct events (*heads* and *tails*). The variable x is a discrete random variable.

We can also have continuous random variables, where x can take any real value (we assume this is a scalar for now). Consider the case of the height of adults in the United States. We will simplify the notation so that instead of writing $p(height = x)$, we will simply use $p(x)$ (as is commonly done). This pdf can be modeled using a Gaussian (or normal) distribution as:

$$p(x) = \frac{1}{\sqrt{2\pi\sigma^2}}e^{-\frac{(x-\mu)^2}{2\sigma^2}}.$$

This is the famous bell curve, plotted in Figure 8.4 (middle). This distribution has a mean μ and a standard deviation of σ. For adult males in the United States, in the case

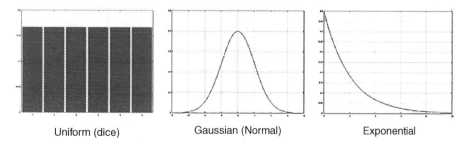

Uniform (dice)   Gaussian (Normal)   Exponential

Figure 8.4 Three different probability density functions. *Left:* Uniform distribution for a fair six-sided die. *Middle:* Gaussian or normal distribution. *Right:* Exponential distribution.

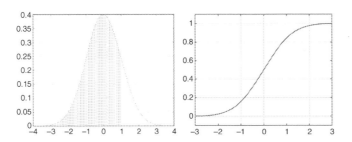

Figure 8.5   The cumulative distribution function for a Gaussian distribution. This allows one to compute the probability that our event has a value smaller than *x*. *Left:* Illustration of the cdf as the shaded area (integral) under the distribution. *Right:* A plot of the actual cdf.

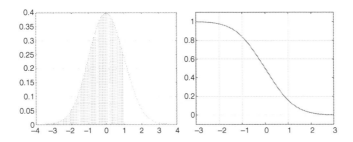

Figure 8.6   The survival function for a Gaussian distribution. This is the opposite of the cumulative distribution and allows one to compute the probability that our event has value greater than *x*. *Left:* Illustration of the survival function as the shaded area under the distribution. *Right:* A plot of the actual survival function.

of height, the mean is approximately 1.75 m and the standard deviation is about 10 cm. The Gaussian distribution is probably the most commonly used probabilistic model as it is easy and convenient to apply to many situations.[4]

In general, to create a pdf like this, we sample the population of interest (i.e. adults in the United States), measure their heights, compute the mean and the standard deviation, and then plug these numbers into the formula. A sample use of this type of model might be to estimate the appropriate height of the doors for an office building. We might specify that we need the doors tall enough that 99 percent of all the employees can safely walk through the door without bending down.[5] We can go ahead and use this model to compute this probability as the following integral:

$$p(x \leq h) = \int_0^h p(x)dx.$$

This integral is also known as the cumulative distribution function (cdf), shown in Figure 8.5. The opposite of this is the so-called survival function, which is the integral that is used to compute $p(x \geq h)$. This function measures the probability that our variable will be greater than a value (e.g. h in this case). It is shown in Figure 8.6. The term survival comes from its use in estimating life expectancy – for example, if x is the lifetime of an individual, the survival function measures the probability that somebody will live longer than h years, whereas the cdf measures the probability that a person will live for fewer than h years.

Multivariate Probability Density Functions The natural next step is to formulate the pdf for two random variables. This is known as the joint distribution. For example, consider the case of the distribution of height and weight. We can use a multivariate Gaussian model for this purpose. First, define the vector $\mathbf{x} = [\text{height}, \text{weight}]$. Then, we can write the multivariate pdf as:

$$p(\mathbf{x}) = \frac{1}{\sqrt{2\pi |\Sigma|^2}} e^{-(\mathbf{x}-\mu)^t \Sigma^{-1}(\mathbf{x}-\mu)}.$$

Here, Σ is the covariance matrix of the two variables height and weight.

Conditional Probability Density Function We are now moving from a function of the form $p(x)$ to one with the form $p(y|x)$. This last function is interpreted as the probability of y given x (i.e. x has already occurred). If we return to the height/weight example from the previous section, the multivariate density function is also the joint density function $p(weight, height)$. An example of a conditional probability is the pdf for height for the case in which the weight is known (e.g. 50 Kg or 110 Lb). This can be expressed as $p(height|weight = 50\,\text{Kg})$. One would guess that this distribution has a mean significantly lower than 1.75 m! Shorter people tend to be less heavy.

Bayes' Rule A key relationship in probability theory is Bayes' rule, which can be written in one of two forms:

$$p(a|b) = p(a \text{ and } b)/p(b),$$
$$p(a|b) = p(b|a) \times p(a)/p(b).$$

This equation specifies the relationship between the marginal probabilities $p(a)$, $p(b)$ (the marginal probability is the pdf for a single variable in the case when we have multiple random variables), the conditional probabilities $p(a|b)$, $p(b|a)$, and the joint pdf $p(a \text{ and } b)$.

Conditional Probabilities in Action Let us consider a real-world example in which we have two events: (1) the event S, which is the presence or absence of a particular sign – this is a categorical/discrete variable that takes values *true* or *false*; and (2) the event C, which is whether a particular patient has cancer – this is similarly a *true/false* variable. What we are interested in is the probability estimate: $p(C = \text{true}|S = \text{true})$, that is, how likely it is that the person has cancer if they exhibit the sign (e.g. positive mammogram for breast cancer). The following is what is known (this is a fictional example) about the situation:

- 100 percent of cancer patients have the sign (probability = 1.0 or 100 percent).
- Cancer prevalence in the population is 1 percent (probability = 0.01 or 1 percent).
- 10 percent of patients that do not have cancer present the sign.

Table 8.1 Expected numbers of patients in each category.

Sign	Cancer true	Cancer false	Total
True	10	99	109
False	0	891	891
Total	10	990	1,000

A tabular way of computing this is shown in Table 8.1 for a population of 1,000 subjects.

In this example, 10 subjects have cancer (10/1000 = 0.01) and 990 are healthy. A total of 109 people have signs (all the cancer patients and 10 percent of the healthy patients). Of those 109 people that have signs, 10 are cancer patients, so the probability of having cancer given that one presents the sign is $10/109 \approx 9$ percent. Our test (are signs present?) therefore has a false positive rate of 91 percent, which is not great and would raise serious doubts as to whether the test should be used.

We can also express the same concept algebraically as follows. First, in this discrete *true/false* example, we can use Bayes' rule to express some of these probabilities of having a true sign as (using "t" for *true* and "f" for *false*):

$$p(S = t) = p(S = t | C = t) p(C = t) + p(S = t | C = f) \times p(C = f).$$

If we express this equation in words, we have the interpretation that the probability of having the sign is the sum of two probabilities. The first is the probability of having the sign when the patient actually has the disease. The second is the probability of the erroneous case of a false positive, that is, the sign is present but not the disease (see Section 8.3.4).

Next, we express our knowledge about the disease and test as probabilities:

$$p(S = t | C = t) = 1.0, \quad p(C = t) = 0.01,$$
$$p(S = t | C = f) = 0.1, \quad p(C = f) = 0.99.$$

Our ultimate goal is to estimate $P(C = t | S = t)$. A little algebraic manipulation leads to:

$$p(C = t | S = t) = \frac{p(S = t | C = t) \times p(C = t)}{p(S = t)}$$

$$= \frac{p(S = t | C = t) \times p(C = t)}{p(S = t | C = t) \times p(C = t) + P(S = t | C = f) \times p(C = f)}$$

$$= \frac{1 \times 0.01}{1 \times 0.01 + 0.1 \times 0.99}$$

$$= 0.09$$

$$\approx 9 \text{ percent.}$$

8.3   An Introduction to Statistics

8.3.1   Some Definitions

Statistics is the science of collecting and analyzing large amounts of data from a subset of the population of interest (the sample) for the purpose of making inferences about the whole population. A key challenge in statistics is to evaluate whether a particular measurement (e.g. difference in height between two groups) was obtained by chance or whether it is likely to reflect some underlying reality.

A statistic (singular) is simply a number extracted from a data set. For example, the mean of the set of numbers is a statistic that tells us something about the data set as a whole. Other examples of such "statistics" include the median (the central value), the mode (the most common value), the minimum value, the maximum value, and many others. The choice of a statistic is dependent on the actual task at hand. If one is trying to estimate the amount of pizza to order for a group, for example, the average number of pieces each person is likely to eat may be an appropriate statistic to measure. On the other hand, if one is designing a building in an earthquake zone, the appropriate statistic might be the maximum earthquake magnitude reported in that area in the past 100 years.

An additional useful term is the *sufficient statistic*. A sufficient statistic can be defined as that summary of a data set that gives you all the information about the data set that you need for a specific task. Going back to the pizza example from above, the total number of slices of pizza that is needed to feed a particular group is just as good (leaving taste aside) a piece of information for deciding how much pizza one should order (the task) as the list of how many slices each person in the group requires. The total is a sufficient statistic for this task, and knowing the detailed breakdown for each person does not add anything to the process. In the software process chapters (see Part III), we will describe a number of documents that move us from user needs to system design to software design. These documents are meant to function as *sufficient statistics* of the information collected in the previous part of a process to allow the readers to perform their task (e.g. design the software) as if they had all the information from the discussions that preceded the writing of the document.

The rest of this section describes two techniques from the field of statistics that have a particular bearing on the process of software design and validation. These are: (1) *estimation* – the process by which we compute some numbers from a data set/population to allow us to make decisions (we also examine the case of estimating pdfs); and (2) *detection* – the process by which one decides whether an event has occurred (e.g. whether there is a tumor present in an image). Both of these techniques are commonly used for medical purposes. We will also need them when we design a software validation plan, in which we are interested in, for example, the accuracy of our software for quantifying some parameter or detecting some event. The section concludes with a description of two important elements in statistics, namely the central limit theorem and the law of large numbers, which undergird much of modern statistical practice.

8.3.2 Estimation

The goal of estimation is to use information/measurements from a data set (sample) to make inferences (educated guesses) about the population as a whole. For example, we may need to estimate the typical time that our software takes to perform a certain task on particular hardware. The process for this might go as follows:

1. Run the software multiple times (trials) on the typical hardware and measure the time it takes to complete.
2. Compute the sample mean (the average) time it takes for tasks over the trials.
3. Use the sample mean as an estimate for the population mean.

This sounds simple (and it sometimes is). However, consider the case where we are interested in estimating not the average time but the likelihood that our task will take less than (for example) 0.5 seconds. In this scenario, one approach might be to first estimate the pdf (see Section 8.3.3) of our measurements and then use this to compute the cdf for a value of 0.5 seconds. This might be more efficient in terms of the number of measurements required compared to a second approach in which we first run the algorithm many times and then simply compute the fraction of those times that the algorithm took less than 0.5 seconds to complete. In some ways, this illustrates a fundamental truth in all statistics. Given enough data, the problem is often easy; the hard task (and where the expertise of statisticians becomes extremely useful) is to be able to say something useful given a limited amount of data.[6]

Estimation is often divided between *point* estimation and *interval* estimation. In point estimation, the goal is to produce a single number that is our best guess of what the actual parameter in the population might be. In interval estimation, the goal is to produce two numbers that act as lower and upper bounds for the specific parameter (the confidence interval), with some associated confidence level such as 95 percent, which is the likelihood that the actual parameter falls within the confidence interval.

8.3.3 Estimating Probability Density Functions

We examine here the process of estimating pdfs, though the principles below are equally applicable to the process of estimating a single parameter or confidence interval. Most probability theory assumes that the pdf is known/given. How does one go about learning it in the real world?[7] In general, the strategy, much of which is common to many estimation problems, is as follows:

1. acquire measurements from a "large enough" sample of "representative" situations;
2. use these measurements (and potentially other "prior" knowledge about the problem) to construct the pdf using either "parametric" or "nonparametric" methods.

As in many situations, the common expression that the "devil is in the details" applies here. Let's examine the five terms in quotes above in a little more detail.

Large Enough The sample needs to be large so as to obtain measurements that generalize to the population as a whole. In all these experiments (e.g. clinical trials of a vaccine), what we are really interested in is the ability to predict what happens in the world at large ("the population"), as opposed to the actual smaller group of people that we are measuring from ("the sample"). A sample that is too small does not permit an accurate estimate of the distribution because, in such a case, a small number of bad or idiosyncratic measurements can have a disproportionate impact on the final result. In more technical language, we need a large sample for our estimates to converge.

Representative Simply put, our sample needs to "look like" our general population. If we are performing a drug study, and the sample to be tested is college students, our results will likely *not* generalize to the whole population, as this includes large numbers of significantly older people.[8]

Prior This little word is the source of enormous debates in statistics. The issue is whether one is allowed to use "external" information as part of the estimation (e.g. smoothness of the data and other prior information). The use of prior information is the hallmark of Bayesian statistics.

Parametric One way to estimate a pdf is to assume that it has a known shape with unknown parameters (e.g. Gaussian, Poisson, gamma). We then use measurements to estimate these parameters (e.g. mean and standard deviation). This type of estimation is often simpler, and one needs a smaller data set to get an accurate estimate. The critical question is whether the chosen base distribution (e.g. Gaussian) is appropriate for the problem at hand. Using the wrong distribution can yield false results. A classic parametric method is the so-called method of moments, in which we assume a model distribution (e.g. Gaussian), compute the mean and standard deviation from the sample, and simply plug them into our model (or use them to compute the parameters of the model).

Non-Parametric In those cases where we do not have a good sense of the shape of the pdf, we can use nonparametric techniques to estimate the distribution. The most common techniques involve use of histograms with potentially additional smoothing (e.g. Parzen windows – see the description in Duda and Hart [55]). The advantage of these methods is that they make weaker assumptions about the form of the distributions. The downside is that these methods require significantly more measurements, as we are now estimating a whole function shape as opposed to a small number of parameters.

8.3.4 Detection

Detection aims to answer questions of the form: *Did an event happen?* or *Is something present?* Much of detection theory starts with radar researchers in and immediately

Table 8.2 The four possible outcomes of a binary detection algorithm. This is a combination of whether an event did actually happen and the output of our detector (yes/no) output. The terms "positive" and "negative" refer to the output of our detector. The terms "false" and "true" refer to the correctness of this output.

	Detector = Yes	**Detector = No**
Event = Yes	True positive	False negative
Event = No	False positive	True negative

after the Second World War, with the ultimate goal of detecting the presence of enemy airplanes within a certain distance of the radar apparatus.

In creating a detection system, one typically creates a function/algorithm that takes a set of input measurements $\mathbf{x} = [x_1, x_2, \ldots, x_n]$ and uses it to compute a function $f(\mathbf{x})$ that is a measure[9] of the likelihood of something happening/being present. Then our detector takes the form:

$f(\mathbf{x}) \geq t$ **positive** result: our best guess is that the event happened.

$f(\mathbf{x}) < t$ **negative** result: our best guess is that the event did not happen.

As shown in Table 8.2, there are four possible outcomes of a binary detection algorithm, *true positive, false positive, true negative*, and *false negative*. Given that some find these terms confusing, we find it helpful to remember that they are defined from the perspective of the detector, *not the event*. The words *true* and *false* signify whether our algorithm's result was correct. The terms *positive* and *negative* describe the output of our algorithm. A false positive (also known as a false alarm), for example, is a situation in which our detector suggests that something happened (positive), but in fact the event did not happen and our detector was wrong (false).

Sensitivity and Specificity *Sensitivity* (sometimes called the true positive rate or the recall) is defined as the fraction of actual true events that are detected. If our algorithm were a diagnostic for some disease, it would exhibit high sensitivity if it correctly detected the presence of the disease most of the time. A sensitivity of 100 percent implies that whenever someone has the disease, they are correctly identified as such. Mathematically, this can be written as:

$$\text{Sensitivity} = \frac{\text{True positives}}{\text{Actual true events}} = \frac{\text{True positives}}{\text{True positives} + \text{False negatives}}.$$

Specificity (sometimes called the true negative rate) is defined as the fraction of actual negatives that are correctly identified. Going back to the diagnostic example, our detection algorithm has high specificity if it correctly identifies healthy people most of the time. A specificity of 100 percent implies that our diagnostic never produces a false positive, i.e. it never labels a person as sick when they are in fact healthy. This is defined as:

$$\text{Specificity} = \frac{\text{True negatives}}{\text{Actual false events}} = \frac{\text{True negatives}}{\text{True negatives} + \text{False positives}}.$$

One final term that should be mentioned is the false positive rate. This is defined as $1 - \text{specificity}$.

Please note that a report on a diagnostic or an algorithm that only specifies the sensitivity or the specificity, but not both, is meaningless. One can always, trivially, make one of these quantities 100 percent at the expense of the other. Let us recall our original detector function: $f(\mathbf{x}) \geq t \mapsto$ positive result. If we make t very high, we can ensure that we do not have any positive results. (Positive here is not a value judgment but simply refers to the result of the detector.) In this scenario, both our false positives and true positives are zero, in which case we have *specificity* = 100 percent and *sensitivity* = 0 percent. Similarly, if we set t to be zero (assuming f is always positive), then we will have no negative results, resulting in *sensitivity* = 100 percent (we miss nothing) and *specificity* = 0 percent.

The value of threshold t is specified either theoretically or experimentally, and it depends on our relative tolerance for the two different errors, namely *false positives* and *false negatives*. This depends on the application. In an airport screening situation in which the cost of false negatives is high (i.e. we missed a dangerous object in somebody's bag), our goal is to optimize for sensitivity.[10] This implies that we would like to minimize the number of *false negatives*. In this case, the threshold is likely to be low and we will have many false positives.

Conversely, in a situation where we need to be very certain before we take action,[11] the goal is to minimize *false positives* or to maximize our *sensitivity*. One example, might be a decision-support algorithm used to decide whether high-risk surgery should be performed. In this case, we would need to be very confident that the surgery is likely to be helpful before proceeding. Our threshold, therefore, is likely to be high and we will have many false negatives, that is, situations where we claimed that the surgery would not have been helpful, when in fact it would have been.

ROC Curves A way to visualize the trade-off between sensitivity and specificity is the receiver-operating characteristic (ROC) curve shown in Figure 8.7, where we plot the performance of our algorithm in terms of true positive rate and false positive rate as a function of the threshold. The ROC characterizes the "skill" level of a user or technology so it provide substantially more information compared to only a single operating point. Such curves are used to compare detection algorithms (and by extension binary classification algorithms, which are essentially detection algorithms). One aggregate measure of algorithm quality is the area-under-the-curve (AUC). Superior algorithms tend to have curves closer to the upper-left corner of the plot (both high sensitivity and high specificity).[12] Sometimes, however, one may be interested only in the performance of the algorithm at a specific range (e.g. high specificity or high sensitivity) and not the

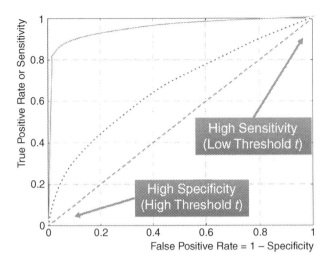

Figure 8.7 Example receiver operating characteristic (ROC) curves. We compare here the performance of three detectors shown in dashed (random guessing – worst), dotted, and solid (best). Each line represents the performance of a single detector algorithm (or user) in terms of the (false positive rate, true positive rate) as the threshold t is changed from zero to a very high value. At $t = \infty$ all algorithms are at (0,0) as our false positive rate is zero. At $t = 0$ all algorithms are at (1,1) as we now have a sensitivity of 1 (we detect everything) and a specificity of 0, resulting in a false positive rate of 1.

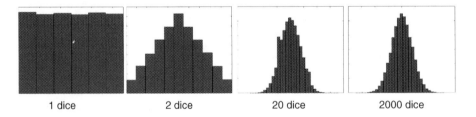

Figure 8.8 An illustration of the central limit theorem. Shown above are the histograms of the mean values obtained after rolling $N = 1, 2, 20, 2{,}000$ dice multiple times. Note that we start with a flat/uniform distribution and move progressively to a Gaussian distribution.

whole ROC curve. The FDA explicitly mentions ROC curves as part of its statistical assessment for diagnostics and biomarkers [75].

8.3.5   Two Important Elements in Statistics

Many common statistical techniques, including the problem of estimating pdfs, rely on two important results from probability theory [247]. These are:[13]

- The central limit theorem (CLT) – see Figure 8.8. This theorem says that under certain conditions, the sum (as well as the mean) of lots of random variables has a normal distribution.[14]

- The law of large numbers, which states that under certain conditions the estimates of parameters of a distribution (e.g. mean) converge to their true value as the sample size increases.

These conditions can sometimes be violated. Consider, for example, a case in which one is interested in estimating the average wealth of a person in the United States. In the usual way, we take a sample from the population, ask them for their net worth, and average the obtained values. This will work for bounded well-behaved variables such as height, but in estimating wealth we may be unlucky and our first subject, randomly, may happen to be a multi-billionaire. Our estimate for the average wealth of a person after 1,000 subjects (even if the last 999 of these have literally zero wealth) will still be over one million dollars, which would clearly be erroneous.

8.4 Machine Learning and Its Complications

The line between machine learning (ML) and traditional statistics is not particularly well defined. Machine learning itself is related to an older field known as pattern recognition [55]. There has been an explosion of interest in this topic, caused for the most part by deep learning techniques arising from the pioneering work of LeCun et al. [152].

For our purposes, we will use the term machine learning to characterize those techniques that use "training data" to learn parameters for an algorithm that is then applied to other data in a subsequent application. For example, consider the case in which we have two variables x and y (e.g. x is heart rate and y is some self-reported measure of stress level). Our goal might be to learn a relationship between x and y so that given a new measurement x_{new} we could predict the stress level y_{new}. In this example, we would first acquire a training set[15] consisting of many measurements of x and y pairs. (This process is known as supervised learning.) From this training set, we would learn a function that predicts y from x – a process known as regression. Then, given a new data point (e.g. in our fictional example a patient would report their heart rate x_{new} to a doctor over the phone), one could predict y_{new} using this prediction model.

A fundamental issue in ML is the problem of algorithm validation and the need to cleanly separate training and evaluation data. When one develops an ML algorithm, the typical steps are:

1. collect a set of labeled data pairs x and y;
2. divide the data into two non-overlapping subsets, a training subset[16] and a test subset;
3. learn the prediction model from the training subset; and
4. evaluate the performance of the model on the test subset.

We note, at once, that the reason for the separate test data set is that we are not interested in how well the model fits the training data, but how well the model generalizes to unseen data (e.g. the test data set and ultimately real patient data when the model is used to help with patient care). The reason for this separation is that most ML models (especially ones using deep learning) have the ability to learn a relationship perfectly. The prediction error in the training set, therefore, might be close to zero, but the model might have poor performance on unseen data. This is known as the generalization problem, and it has a variety of causes, such as (1) insufficient training data; (2) training data that is not rich enough to be representative of the actual situation; or (3) a model that over-learns the idiosyncrasies of the training set (as opposed to the bigger picture), and as such fails on real data.

These are all classical problems in ML, but they are becoming even more pronounced with the advent of deep learning techniques. Deep learning models have millions of parameters that need to be estimated from a training set. Much like proper software engineering techniques, with ML based software, ideally one should have completely independent evaluation (see also Figure 2.2), where not only the test data but also the validation process is performed completely separately from the training. Otherwise, there is a risk of training/test data contamination or model over-training.[17]

The most common ML tasks are regression and classification. In regression tasks, the goal of the model is to predict an actual value (e.g. stress level). In classification, the goal is to categorize the input data into discrete classes (e.g. normal weight, overweight, obese). These are analogous (in some ways) to statistical estimation and detection. We will discuss classification in more detail next, as an example.

8.4.1   Classification

Classification is the process of taking an input data set and assigning to each element a class label (e.g. underweight, normal weight, overweight, obese). In the special case of two-class classification, this is basically equivalent to detection, and in fact many of the same methods (e.g. ROC and AUC) are used to quantify the performance of many classifiers. An example of a linear classifier (one in which the decision boundaries are straight lines) is shown in Figure 8.9. This is a hand-crafted classifier that was designed by expert humans as opposed to being the result of a computer algorithm.

Historically, designing a classifier proceeded in two steps:

1. Extract features from the input element. In the case of BMI (body mass index), the input element is a person and the features are the weight and the height of that person.
2. Design a strategy (classifier) to assign labels to each data point (defined by the features).

Typically, the process has two steps: We first acquire a training data set and then use this to create the classifier. Classifiers can be based on probabilities (see Duda and Hart [55]), or they may directly estimate the decision boundaries. In the latter case, the boundaries may be explicit (e.g. actual boundaries) or implicit (e.g. in the case of nearest-neighbor classifiers).

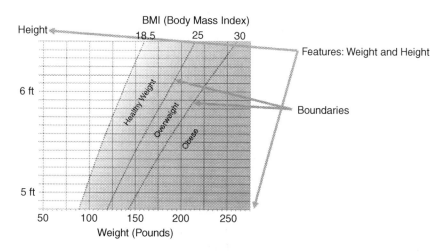

Figure 8.9 A simple "manual" classifier. This figure illustrates the core aspects of a simple classifier. We define our input data as the vector (weight, height) – these are the features of our classifier. Then we define our classes (obese, overweight, healthy weight, and underweight) and define the decision boundaries used to make decisions. The chart is used to define both the BMI score and its use in classifying subjects. In this example, the classes are defined by the boundaries, which are thus perfect – as opposed to the typical case where the boundaries need to be determined. Figure obtained from the National Institute of Digestive Diseases and Kidney (NIDDK).

In these more traditional, feature-based approaches, a significant problem is often caused by having too many features for the size of our training set. There are many techniques to perform dimensionality reduction, such as principal component analysis (PCA).

The major advance in deep learning is the integration of the feature extraction and classification processes. In contrast to traditional techniques (including neural network-based approaches) which require a separate feature extraction step, deep learning methods use "deep" (as measured in layers) neural networks, where the early layers effectively perform feature extraction, and the later layers perform the classification task. By optimizing these tasks together, one can obtain superior performance as the (implicit) features that are learned are, in some sense, optimal for the classification task.

8.5 Are These Results Statistically Significant?

When performing clinical validation of our software, our goal may be to show (1) that the software's performance is acceptable; (2) that our software's performance is better than that of competing technology; and/or (3) that our software's performance is as good as that of some more expensive method. In all of these scenarios, we can perform a study, acquire measurements, and try to show that we have met our criteria for success. A critical question is, however, whether our results (taking the second case where we try to show improvement) are truly better (assuming they are!), not just better due to random effects of the specific sample used (e.g. the actual patients).

This is the classic problem of estimating statistical significance. Consider, for example, the case of testing a new drug. Our claim is that our drug improves (reduces) hospitalization time for some disease. To test this, we first recruit patients and potentially divide them into two groups. The patients in the first group, the treatment group, are given the drug. The patients in the second group, the control or placebo group, are given an identical-looking alternative to the drug. If the drug is tested in pill form, for example, the placebo would be a pill that looks identical to the "real thing" but, instead, contains an inactive substance such as starch or sugar.

While there are all sorts of complicated statistics that one could perform on the data, the first-level analysis is to compute what is known as the effect size [169]. For example, we may begin by computing the mean hospitalization time for the treatment group (we will call this μ_t) and the mean hospitalization time for the control group (μ_c). We could then define our effect size as, for example, $\mu_t - \mu_c$. This is the average real-world impact of our drug.[18] If this is small, then statistics often do not matter. If the effect size is large enough for it to be interesting, then we proceed to ask the second question: Is this effect real or was it obtained by chance? In more formal scientific vocabulary, the question is phrased as "Are the results statistically significant?"

8.5.1 Statistical Hypothesis Testing

Statistical hypothesis testing is a technique for estimating statistical significance. The setup is a little counterintuitive, but hopefully can be explained by continuing with our drug study example.

The first (and critical) step in hypothesis testing is to state what is known as the "null" hypothesis. This is almost always the opposite of what we are trying to show. In our case, the null hypothesis might be "the drug has no significant effect." The next step is to use our measurements to reject this hypothesis, that is, to demonstrate that, assuming that the null hypothesis is correct, the probability of obtaining the measurements we obtained (or larger[19]) is below a certain significance level (often denoted as α) such as 5 or 1 percent. If this is the case, we reject the null hypothesis as unlikely to be true, and by extension prove the opposite: "our drug has a statistically significant effect."

The computation proceeds as follows. First, we assume that there is no effect. This means that (some measure of) the hospitalization time of our control group and our treatment group are supposed to have the same probability distribution (i.e. it is the same). Next, we learn the pdf of hospitalization time from our data – see Section 8.3.3 for techniques for doing this. Next, we compute the probability of obtaining measurements at least as different (think of the survival function – see Figure 8.6) between the two groups. If this probability is small (e.g. below 5 or 1 percent), we can then conclude that the (null) hypothesis that the two groups are the same can be rejected, and by extension that our data shows statistically significant differences.

Please note that if we fail to reject the null hypothesis, that is, our probability is > 5 percent, this does not constitute proof that the data from the groups is the same. All it means is that we have failed to prove that they are different. Thus, this technique cannot be used to show equivalence/similarity, only differences. This is a common mistake that one sees in the literature.

Even though it is probably the most commonly used statistical technique in scientific studies, there are many, many problems with hypothesis testing. The most important one is the appropriate selection of a null hypothesis and the correct assumptions as to how to test for it. For example, in computing the pdf above, most studies make the assumption that the measured data derives from a population that has an underlying Gaussian distribution, and as such uses the t-distribution for the pdf of survival time.[20] Unfortunately, this is often done blindly, without checking that the t-distribution model is appropriate. If the model is not appropriate for our data, then the results should not be trusted.

8.5.2 Some Commonly Used Terms

In a hypothesis testing setup, there are two types of errors:

- Type I error ↦ false positive. In this case, our statistical procedure rejects the null hypothesis even though there was no effect.
- Type II error ↦ false negative. In this case, our statistical procedure fails to reject the null hypothesis even though there is an actual effect.

Two other terms that are used are:

- α (alpha) – the significance level at which we reject the null hypothesis, for example 0.05 (5 percent) or 0.01.[21] This is the probability of a Type I error.
- β (beta) – the probability of a Type II error.
- *power* $= 1 - \beta$, the probability that a test will correctly reject a false null hypothesis.

Based on these, we can characterize a statistical procedure as having "high statistical power" if there is a small risk of committing type II errors. This means that if there is an effect, we are highly likely to detect it. In most experimental trial designs, the sample size is estimated to ensure appropriate power. Increasing the sample size N improves the ability of the test to detect a difference, as the uncertainty typically goes down roughly as $\sqrt{N}$. Typically, one first chooses a certain power level (e.g. 0.80), and then based on one's knowledge of the uncertainty (standard deviations) from prior observations, computes the sample size that is required.

8.6 Importance of Randomization in Clinical Trials

The discussion of statistical significance leads us naturally to the topic of randomized clinical trials. (We discuss clinical trials in more detail in Section 15.2; our focus here is on the effects of randomization.) These are structured experiments in which we try to demonstrate that a proposed intervention[22] has a significant positive effect on a certain population that is not observed simply by chance. In randomized clinical trials, participants are assigned to (for example) the treatment or the control group "by chance." This allows us to eliminate hidden biases that can arise from which types of people would volunteer for treatment. (Naturally an effort is made to balance the groups in terms of sex, age, etc.)

Failure to randomize can have significant negative (and strange) effects. Judea Pearl, in his book *The Book of Why* [205], illustrates the well-known Simpson's paradox, using an example case of a drug that simultaneously appears to be "bad for females, bad for males and good for all people regardless of sex." This erroneous result was caused by a lack of randomization in the sample. This example derives from a trial for a drug that was designed to prevent heart attacks. In that trial, the study participants *were given a choice* as to whether they took the drug or not; they were not randomly assigned to each group. Following the experiment, the results were:

- Female subjects: The fraction of subjects that had a heart attack increased from 1/20 (5 percent) to 3/40 (8 percent) for women who took the drug. Conclusion: *Adverse effect in females.*
- Male subjects: The fraction that had a heart attack increased from 12/40 (30 percent) in men who did not take the drug to 8/20 (40 percent) in those who did. Conclusion: *Adverse effect in males.*
- Total population: The fraction that had a heart attack *decreased* from 22 percent to 18 percent. Conclusion: *Beneficial effect in people.*

The problem was, as alluded to earlier, failure to balance the groups by sex. Male participants had more heart attacks (33 percent) than female participants (7 percent). Since female participants volunteered to take the drug in larger numbers, this created the false result that overall the drug actually had a positive effect when in fact it did not.

In this case, one could argue that the paradoxical result is simply a lack of group balancing and not a failure of randomization. This is true for this specific case, where the difference in the groups (sex) is obvious. There are many other subject characteristics, however, that are impossible to observe but may correlate both with the subject's willingness to take an experimental drug and their response to it. Perhaps risk aversion (i.e. unwillingness to try a new drug) correlates with outcome. Randomization eliminates this and other hidden biases.

Implications for Software Validation One might ask, why is this relevant to software validation?[23] Consider the case where our claim for our software is that it leads to a 20 percent reduction in misdiagnosis by radiologists reading MRIs. Let us further assume that, in reality, there is no benefit to our software – that is, our claim is false. Naturally, we do not know that our claim is false prior to performing measurements. To test our claim, we proceed to recruit two groups of radiologists and ask them to read known images that both contain and do not contain both cancers. We then compute their accuracy. Group 1 uses their existing setup, and Group 2 uses our software. If we somehow manage to get all the *good* radiologists into Group 2, we might be able to (erroneously) demonstrate an improvement, as the "better" radiologists will have better scores. For example, if we ask radiologists to volunteer for the method they would like to use, it may be that *good* radiologists are more willing to experiment and may end up disproportionately in Group 2, thus showing an erroneous improvement. The standard solution to this problem is to randomly assign subjects to groups.

8.7 Conclusions

The take-home lesson from this chapter is that the development of medical software requires an understanding of mathematical techniques from the fields of probability theory, statistics, and ML. This is not only the case for those parts of the software that directly use such techniques (e.g. the use of supervised deep learning algorithms for part of the functionality of the software), but also for the design of validation strategies where issues such as estimating sample properties, determining statistical significance, and accounting for the need to randomize study subjects are critical.

8.8 Summary

- The ability to estimate parameters from a data set is critically dependent on both the size and the diversity of the population from which measurements are obtained. Small data sets cannot accurately capture the true variability of the population.
- An understanding of the underlying assumptions behind many commonly used statistical methods is important to enable the software developer to avoid obvious pitfalls (e.g. use of inapplicable methods, misunderstanding of what statistical significance means, blind application of ML techniques).
- Despite current popular beliefs, the data will not "tell you everything." Sometimes new situations arise (e.g. the COVID-19 pandemic) which make predictions based on old data inapplicable. Hence, the use of purely data-driven techniques can also sometimes run into problems regardless of how "big" the data sets are.
- Randomization of study subjects (whether the subjects are the user of the software or actual patients) is required to avoid hidden biases in validation studies.

RECOMMENDED READING AND RESOURCES

A good introduction to probability theory can be found in the popular textbook by Papoulis and Pillai:

A. Papoulis and S. U. Pillai. *Probability, Random Variables, and Stochastic Processes*, 4th ed. McGraw Hill, 2002.

The classic textbook in pattern recognition is by Duda and Hart. This book in many ways anticipated many of the subsequent developments in the field of machine learning:

R. Duda and P. Hart. *Pattern Classification and Scene Analysis*. Wiley, 1973.

A very readable introduction to the topics of artificial intelligence and machine learning can be found in:

M. Mitchell. *Artificial Intelligence: A Guide for Thinking Humans*. Farrar, Straus and Giroux, 2019.

A more technical introduction to the topic of deep learning can be found in the following book, which is also freely available online:

I. Goodfellow, Y. Bengio, and A. Courville. *Deep Learning*. MIT Press, 2016. www.deeplearningbook.org

A recent review by Esteva et al. summarizes the current state of the use of deep learning in healthcare:

A. Esteva, A. Robicquet, B. Ramsundar, et al. A guide to deep learning in healthcare. *Nat Med*, 25(1):24–29, 2019.

Nassim Nicholas Taleb's many books discuss the issues with fat-tailed distributions and how these affect our ability to predict events. The classic one is:

N.N. Taleb. *The Black Swan*, 2nd ed. Random House, 2010.

Finally, the history of the development of the *t*-distribution is told very nicely in the following online article:

D. Kopf. The Guinness brewer who revolutionized statistics. https://priceonomics.com/the-guinness-brewer-who-revolutionized-statistics.

NOTES

1. These comments echo material that is presented in a far more interesting (and highly entertaining) manner in the many works of Nicholas Taleb (e.g. *The Black Swan* [246]). Naturally, the problem is that what was thought to be impossible based on prior observations (big data) and modeling turned out not to be impossible after all.

2. All probability theory seems to have originated for the needs of gambling. One often sees words like "fair" coin appear in this vocabulary, as cheating/malfeasance seem to be common in that domain.

3. Placebo: In many treatment studies there is a well-known phenomenon known as the placebo effect, where patients improve if told they are been given a drug, even though

they are really receiving an inactive, harmless substance. It is common in drug studies to use placebo as a control condition to ensure that any improvements as seen are actually a function of the drug studied and not purely psychological benefit that occurs from feeling that one is receiving treatment. See also Section 8.5.

4. It is often used in cases where it is inappropriate. Almost any time a characterization of some data set is reduced to computing the mean and the standard deviation, there is the implicit assumption that we have a Gaussian pdf or a related pdf such as the t-distribution. The Gaussian is a thin-tailed distribution [247]. The probability of a normally distributed variable having a value more than three standard deviations away from the mean is practically zero. This is very convenient for many statistical estimation tasks, but often an unrealistic model of many actual situations.

5. Unless, of course, this is a building housing a professional basketball team.

6. Consider the case of pre-election opinion polls. There are lots of tricks used by pollsters to reweigh their results to account for differences in sample and population properties (e.g. fraction of college graduates) to get a better estimate. If, however, one could sample one million people instead of 1,000, the raw results might be just as good. This recalls the old maxim that, to paraphrase, an engineer is somebody who can do for $1 what everybody else can do for $10.

7. This is required for any type of statistical significance testing, for example. When using turnkey statistical toolboxes the estimation of such pdfs is done implicitly by assuming certain parametric forms (e.g. Gaussian). There is no escaping this process though. If the underlying pdfs have fat tails [247], the sample size requirements can be dramatically larger than for thin-tailed distributions such as the Gaussian. Using the wrong model for the pdf can lead to erroneous results.

8. The same considerations apply to other variables such as sex, IQ, height, weight, etc. These variables may or may not be applicable, depending on what one is trying to measure. Sample size and diversity are critical when using some of the hypothesis-testing techniques, as is common in many scientific studies, as when applying these techniques we are implicitly estimating the pdfs for the outcome variable of interest.

9. This need not be a formal probabilistic description, so long as it increases monotonically as the likelihood of the event happening increases. $f(x)$ could be as simple as a linear function, or as complex as a deep neural network.

10. Directly after September 11, 2001, one of the authors was traveling through a German airport where they inspected every bag manually. This was a case of optimizing for sensitivity at the expense of specificity (i.e. all the bags were treated as suspicious and opened for manual inspection and almost none of them contained any dangerous materials).

11. This is how the justice system in the "Western world" has traditionally (at least in theory) operated, where the standard for declaring somebody guilty of a crime is high.

12. There are other methods such as LROC (localization ROC) and FROC (free-response ROC) which are used in cases where one is not simply interested in the binary detection case (e.g. does the patient have cancer or not?) but also the localization of the cancer

is important. Localization is, for example, an important component in computer-aided diagnosis based on medical images. See He et al. [106] for a good overview.

13. This fact is often not understood/known to the users of these techniques that apply them blindly!

14. In some distributions, such as the Cauchy distribution, the mean does not exist! Clearly the CLT does not apply to these.

15. It is becoming increasingly accepted that the construction of training sets is one of the most critical tasks in ML – see Section 9.3. The work needed to create such training sets is often labeled as "data janitoring" – see, for example, this article that appeared in the *New York Times* [160].

16. This is further subdivided into a training subset and an evaluation subset. The latter is used for initial evaluation, which allows the tweaking of the algorithm by the designer.

17. In many scientific papers, deep learning models are validated using cross-validation techniques to compensate for the lack of truly large training data sets. In the extreme scenario known as "leave one out," given a data set of size N, one learns many models each based on N – 1 pairs and tests on the remaining pair. This process is repeated N times and the average performance score is computed. A more advanced version is k-fold cross-validation. While this process is technically correct as the training data and the test data are not mixed, in practice the same data is used to tune the algorithm (or test multiple algorithms), thus introducing contamination via the researcher tuning the model manually.

18. There are many other statistics one could compute. For example, one might discover that the drug only has an effect on the most sick patients. In such a scenario, the mean hospitalization time over all patients may not show any difference.

19. Assume that we observe a difference in hospitalization equal to three weeks. What we need to compute in this case is the probability of observing a difference of at least three weeks, assuming that the null hypothesis is correct. Thus, the appropriate function to compute is the survival function of the null distribution.

20. The development of the t-distribution comes from the work of William S. Gosset [148]. This work was performed at the Guinness brewery in Dublin, Ireland as part of attempts to standardize the taste of beer. The key mathematical result of this work was that if the original data is normally distributed, samples from this data are distributed according to the so-called t-distribution. As the sample size becomes larger, the t-distribution becomes more and more similar to a Gaussian. This is the origin of terms such as "t-tests."

21. The infamous 0.05 threshold that for many years determined (and still sometimes does) whether a study is published or not (i.e. whether the results are considered significant) comes from a statement by Fisher: "The value for which P = .05, or 1 in 20 ... it is convenient to take this point as a limit in judging whether a deviation is to be considered significant or not. Deviations exceeding twice the standard deviation are thus formally regarded as significant" [94]. Please note the implicit assumption in this statement is that the data has a Gaussian distribution.

22. This could be a drug, a treatment procedure, or the use of a piece of software.

23. In the emerging cases where software is used as a therapeutic, the relationship is obvious.

9 Topics in Software Engineering

INTRODUCTION

This chapter provides some basic background on selected topics in software engineering. This should be useful for those readers whose background is primarily in basic science and engineering, who may have not been previously exposed to this type of material. This chapter is meant to complement the introduction to mathematical topics presented in Chapter 8. We begin with a brief overview of software engineering (Section 9.1). Next, we discuss the software life cycle (Section 9.2), the organizing foundation of any software project. Following this, we present issues related to the use of artificial intelligence/machine learning (AI/ML) techniques (Section 9.3) in software, in particular the centrality of data in this process. Next, we discuss modern computing environments such as cloud-based infrastructures and mobile devices (Section 9.4). The chapter concludes with a description of two core software engineering topics: software testing (Section 9.5) and source code management (Section 9.6).

9.1 Software Engineering Overview

In Part III of this book, we will present an example of a medical software life cycle process; this chapter provides the necessary background in software engineering that is required for a full appreciation of this process. High-quality software is not created by simply having a programmer sit (or stand) in front of a computer and code. As discussed in Chapter 4, one must first have an organizational structure and processes that support the software engineering effort. In addition, complex software development follows a life cycle. The IEC 62304 standard [119] defines the software development life cycle model as a "conceptual structure spanning the life of the software from definition of its requirements to its release for manufacturing" [119]. There are a number of possible software life cycles, including the classic, and much-maligned, Waterfall life cycle and more modern (but not always better) Agile techniques.

In a large project, code must be appropriately managed and tested. Source code management is a critical component of any software project. In a project of any significant size, the work of producing the actual code of our software will be shared by multiple developers, so we need processes to ensure that "the pieces fit together." The process for source code management should be explicitly covered in an organization's quality management system (QMS) (see Chapter 4). In addition, code must be tested.

While we will discuss testing in Chapter 14, in this chapter we discuss the basics of what software testing is and how it is used.

The software engineering process becomes more complicated when our software involves the use of AI/ML modules, as the training data, though a critical component of the software, is not part of the code and must be handled with care. The discussion in this chapter, which is based on emerging regulatory guidance documents [53, 189], provides the necessary foundations for the discussion of the implications of the use of AI/ML techniques throughout Part III, and in particular the description of software validation in Chapter 15.

9.2 Software Life Cycles

We introduce here the concept of software life cycles. Most complex software projects require structure and organization and involve multiple people working as a team over a significant period of time. This complex process requires an organizing principle, or model.[1]

In Chapter 10, we will present a sample life cycle process for medical software development that is based on both regulatory guidance and industry standards (in particular IEC 62304). As an introduction to this topic, IEC 62304 [119] defines three basic types of life cycles. These are:

Waterfall: This is a "once-through" strategy. The requirements are all defined at the beginning. This cycle uses a single development cycle consisting of "determine customer needs, define requirements, design the system, implement the system, test, fix, and deliver."

Incremental: This strategy differs from the Waterfall in that following the definition of system requirements, the development is performed "in a sequence of builds. The first build incorporates part of the planned capabilities, the next build adds more capabilities, and so on, until the system is complete."

Evolutionary: This takes the incremental strategy a step further in that the customer needs and system requirements are also revisited and refined during each build step. The underlying assumption made when using this strategy is that "the user need is not fully understood and all requirements cannot be defined up front."

The standard categorizes the three types of models using three defining characteristics:

1. *Define all requirements first?* Waterfall and incremental (yes), evolutionary (no).
2. *Multiple development cycles?* Waterfall (no), incremental and evolutionary (yes).
3. *Distribute interim software?*[2] Waterfall (no), incremental (maybe), evolutionary (yes).

While most regulatory documents and standards are written as if one is using a 'Waterfall'-style model, it must be stated explicitly that the FDA does not prescribe

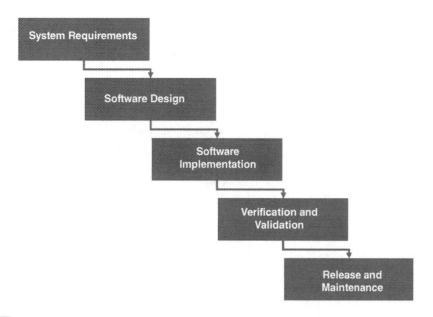

Figure 9.1 The Waterfall software life cycle model. This is the (often justly criticized) traditional software development model. We start by defining the requirements for the project and proceed to the actual software design, the software implementation, verification and validation, and finally release and maintenance.

any specific life cycle but simply that one must have one (the same applies to industry standards). In fact, as Kappe points out [143], the FDA guidance explicitly acknowledges that the Waterfall model does not match reality by stating that: "Software requirements are typically stated in functional terms and are defined, refined, and updated as a development project progresses" [70]. Hence, while many textbooks (and regulatory guidance documents) are written as if one were using this model, in practice this is an idealization and is modified in many ways. The key is to make these modifications in a controlled manner, such that the project stays in-sync, i.e. the software design is the one that gets implemented and not some arbitrary modification made by the programmer(s) without informing software designers and users.

9.2.1 The Waterfall Model

The original cycle is the so-called Waterfall model (see Figure 9.1), first described by Royce in 1970 [223]. This represents an idealized sequential process. We begin by establishing the software requirements based on discovering the needs of our users. The next step is to create a complete design for the software, at the appropriate level of detail. This design is then implemented as code. When the implementation is completed, one proceeds to verification (testing that the software works correctly according to the requirements). Finally, the software is released to the user and enters maintenance mode, where bugs are fixed and new features added (in a process that is almost a recursive repetition of the Waterfall model itself). The name Waterfall comes from the fact that the process proceeds serially from one level to another, much like water going down a Waterfall, with no "cyclical" looping from one step to the next.

The Waterfall model presents a highly idealized situation and has many drawbacks. Two stand out in particular: (1) The serial aspect of the process is often impossible to adhere to in practice, as we often discover issues in later steps of a process that affect prior steps. For example, during the implementation phase, we may discover problems that make the design of the software impossible to implement, and which could not have been anticipated before. And (2) the Waterfall is an all-or-nothing process, meaning that there are no "in-between" versions of the software available prior to the completion of the entire package. This is sometimes unacceptable, given that the process is often time-consuming. It is often useful to have functional but incomplete prototypes. These permit better testing and can be invaluable in identifying issues with the design and even the requirements. In the absence of functional prototypes, such design/requirements issues may only become apparent when the software is complete.[3]

There are some common modifications of the Waterfall model, such as the Sashimi model [245]. This model is an attempt to overcome the fact that the theoretical Waterfall process is not usable in practice. It assumes that the various phases (e.g. Requirements, Design, Implementation, Testing) are overlapping and that explicit iteration between the different phases (and teams involved) is needed.

Another related approach is the so-called "prototyping" model, where essentially we build an initial version of the system "all the way through" to establish the process (requirements, design, etc.) and then, using the knowledge learned from this, go back and follow the process again to create the "stable" version. This is often the case when the initial system derives from research prototypes produced in academic settings.

9.2.2   Incremental and Evolutionary Models: Agile

In these types of models, the functionality of the system is divided into small blocks that can be implemented quickly. The process moves from the most important functionality [224] on to the rest. There are a variety of techniques that fall into this area. One will hear terms such as "Test-driven" [267], "Agile" [18], "Extreme Programming" [17], and "Scrum" [225]. We will use the term "Agile" as a loose definition of all these approaches. The defining statement on Agile software development is the "Agile Manifesto (2001)" [18]. To quote this in full:

> We are uncovering better ways of developing software by doing it and helping others do it. Through this work we have come to value:
>
> *Individuals and interactions* over processes and tools
> *Working software* over comprehensive documentation
> *Customer collaboration* over contract negotiation
> *Responding to change* over following a plan
>
> That is, while there is value in the items on the right, we value the items on the left more.

An initial impression might be that the Agile Manifesto is incompatible with the medical software regulatory process. Statements such as "over following a plan" or "over comprehensive documentation" appear to run counter[4] to regulatory guidance for medical software development, with its explicit emphasis on plans and documents. However, these methods have become increasingly accepted for software development in the medical domain, with some constraints, as we will discuss next. Our discussion will be guided by the AAMI Technical Information Report titled *Guidance on the Use of AGILE Practices in the Development of Medical Device Software* [1].

A Review of Agile Life Cycles The goal of AAMI TIR 45 [1], to address "questions from both manufacturers and regulators as to whether (or which) Agile practices are appropriate for developing medical device software." As the TIR states, Agile can bring value to medical device software and can be adapted to the unique needs of this domain. The document then goes on to make suggestions as to how to implement Agile methods for medical software. Before discussing these adaptations, we will review the Agile process following the discussion in this document. When using Agile, a project is organized hierarchically into the following:

Release: A project consists of a number of releases. Each release results in a working piece of software that could be made available for user testing – though typically not all releases are. (Typical duration: several months.)

Increment: A release consists of a number of increments – activities that combined create useful functionality. (Typical duration: 1–4 weeks.)

Story: An increment consists of a number of stories. A story is a simple description of a feature from the perspective of the person who needs it (e.g. a user). It follows a who/what/why template. As an example, a story might be *"As a user I would like to be able to measure tumor volume from MR images so as to assess disease severity."* (Typical duration: a few days.)

Finally, we define here the term *backlog*, which is the list of stories for which functionality still needs to be implemented.

Adapting Agile for Medical Software We will not present here a comprehensive description. Those interested in applying Agile for this purpose should consult the AAMI TIR [1]. We will restrict our discussion to aspects that we feel are particularly worth highlighting.

1. Documentation: The Agile manifesto values "Working software over comprehensive documentation." This, however, needs to be reconciled with the fact that both the regulators and the standards require documentation to demonstrate that an appropriate process was followed. As the TIR states, "Saying 'we are Agile, we don't need design documentation' is not a defendable position in the medical device software world ... a common principle in the medical device software world: 'If it isn't documented, it didn't happen.' " Hence documentation is needed, but perhaps the intuition behind this value "software over documentation" is to focus on what documentation actually adds value and not to over-emphasize documentation at the expense of the actual software.

2. Mapping the Agile process to IEC 62304 activities: The TIR provides mapping between the life cycle described in IEC 62304 (see Figure 10.2, which is implicitly based on the Waterfall model), and an "Agile" life cycle. The goal is to demonstrate how Agile can be used for medical software development. The recommendations are as follows:

- At the "project" high level we perform initial development planning, requirements analysis, and an initial architectural design for the whole project.
- For each "release," we first perform initial planning. Then we move into a number of increments (more next), concluding with integration and system-level testing and software release.
- For each "increment" we follow a reduced version of the "release" process. Here we perform initial planning, move to working on a number of stories, and then perform a final integration and system-level testing.
- At the "story" level, we follow more or less the entire IEC 62304 development life cycle (planning, requirements, architectural design, implementation, testing) excluding the software release stage.

One of the challenges of adopting an Agile life cycle is the accessibility of stakeholders. For example, an isolated company which does not have easy access to its users may not be able to implement a process where constant interaction with the users is required. Sometimes this constant interaction can also introduce problems, as there will be the temptation to keep changing the requirements – see the case of the Healthcare.gov website described in the vignette presented in Chapter 19.

On a more positive note, in a conference paper titled *Adopting Agile in an FDA Regulated Environment* [212] by Rasmussen et al. there is a case study from the Abbot Diagnostics Division's adoption of Agile techniques for their m2000 project. The process had the following four key principles:

1. fixed and short duration iterations (4–10 weeks) delivering functional software;
2. automated continuous build, unit test, and integration;
3. daily team meeting providing high visibility of status and promoting concept of collective ownership;
4. retrospective assessment at conclusion of each iteration by key stakeholders.

The company found that they could reduce the overall project duration by 20–30 percent using this type of approach.

9.3   Software Engineering with AI/ML Modules

We discuss here challenges that are introduced in the software engineering process by the increasing use of AI/ML modules. Our focus will be on the use of supervised ML techniques and in particular deep learning methods (see Section 8.4). Our comments, however, are generally applicable to other data-driven techniques. As primary sources, we will follow the document DIN SPEC 92001-1:2019 [53] from the German Institute

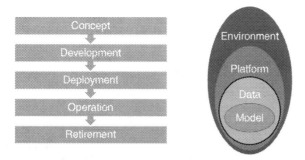

Figure 9.2 AI quality metamodel. This figure summarizes the life cycle of a software module that uses AI/ML techniques. Adapted from DIN SPEC 92001-1:2019 [53].

for Standardization and the AI Guidance from the NMPA (the Chinese regulator) [189]. The discussion in the *Regulatory Guidelines for Software Medical Devices – A Life Cycle Approach* [234] document from the Health Sciences Authority of Singapore is also very useful. Other relevant material is listed in Section 1.3.

Figure 9.2 summarizes the metamodel proposed in DIN SPEC 92001-1:2019 for the development of software modules that use AI/ML techniques. The central portion of this figure is similar to the standard software Waterfall life cycle. Critically, though, the figure highlights the importance of other aspects, such as data and external criteria. DIN SPEC 92001-1 also discusses three "Quality Pillars" for AI modules (see also the leftmost module in Figure 9.2). These are:

1. **Functionality and performance** – measurements of the degree to which our AI module "fulfills its intended task under stated conditions." This leads to a requirement that one must specify explicitly both the quality metrics and the conditions under which our software will operate (e.g. the type of input data, which could be, for example, T1-weighted MRI data acquired from patients between the ages of 20 and 60 with resolution at least 1 mm at a field strength of 3T).
2. **Robustness** – a measure of the ability of our module to handle "erroneous, noisy, unknown and adversarial" input data. This is part of risk management, and we will discuss it in the next chapter (Section 5.2.4).
3. **Comprehensibility** – this is defined as "the degree to which a stakeholder with defined needs can understand the causes of an AI module's output." AI modules are often black boxes with limited interpretability, that is, the user often has no idea why the module's answer is what it is. It is often helpful for such modules to output not only a result but also some form of visualization/explanation that highlights the key factors (e.g. part of the image) that drove the particular output. There is ongoing movement towards "Explainable AI" [226, 254].

From a software engineering perspective, the major paradigm shift when using AI/ML modules is that critical components of the software engineering process *do not live within the actual code of the software itself.* These include the training code (i.e. the code used to train the model) and the training data itself. While other types of

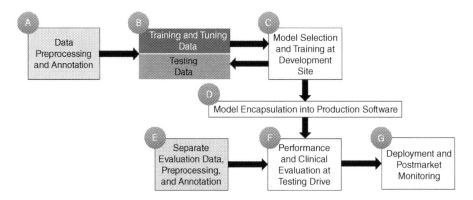

Figure 9.3   Development and deployment life cycle for AI/ML modules. This figure summarizes the process of development and deployment of AI/ML models. A possible complication not shown in the figure is that the code used for preprocessing the images should probably be included in the final software (Box D), and, correspondingly, the data used for the evaluation (Box E) should not be preprocessed as the preprocessing should be done within the production software. This is a result of the expectation that our software will take raw data as inputs when operating in a production (clinical) environment. Adapted from a guidance document produced by the Health Sciences Authority of Singapore [111].

software use external resources (e.g. images to be used as logos, text databases, etc.), these are often secondary in importance and often play cosmetic roles. By contrast, in AI/ML modules these external resources are often where the actual value of the software lies [8]. We discuss this process next in order of importance, starting with *the data*. A flowchart is shown in Figure 9.3.

9.3.1   Training Data Management

The importance of training sets in deep learning algorithms cannot be overemphasized. The size and construction of training and validation data sets is often the differentiating factor between similar applications. By contrast, both low-level deep learning algorithms (e.g. optimizers) and higher-level deep learning models are gradually being standardized and are easily available as part of popular open-source toolkits. The most popular such toolkits are Google's TensorFlow [250] and Facebook's PyTorch [210], which are either used directly or through domain-specific toolkits, such as MONAI [179] for medical image analysis. For many tasks there is also a convergence toward a particular class of deep learning model. For example, for medical image segmentation tasks there seems to be a convergence toward variants of the U-Net Model [136, 217].

The data is often the most critical component of the process. As stated in DIN SPEC 92001-1:2019 [53]:

the data used by an AI module plays a significant role in its performance and quality. Due to the increasing amount and use of large data sources, it is particularly important to address issues of representability, availability, and quality of data. In the context of ML, these aspects can have a significant influence on achieving the intended outcomes. Therefore, data is one of the major topics when reasoning about the requirements of AI modules.

In most software engineering processes, the handling of a significant amount of data comes at the end of the process, when we perform verification and validation. Hence, the management of data is often the domain of software testers/evaluators (as opposed to developers), who are often given just small sample data sets to enable their initial testing but not much more. This is emphatically not the case in AI/ML. The training data is critical to the development process, and the management of it requires significant care and documentation from the beginning of the process.[5]

There are different aspects of data management, as discussed in the NMPA guidance document on AI [189]). These are:

1. data acquisition;
2. data preprocessing;
3. data annotation[6];
4. data set construction.

Data acquisition in this context is often literally obtaining data from an outside source such as a hospital. Such data needs to be quality controlled, appropriately de-identified to comply with privacy regulations (see Section 3.4), and appropriately indexed in such a way that key metadata can be easily found (e.g. age, sex, race). The processes for both quality control and anonymization may involve external software, which should be treated in the same way as any other development tool.

Often, our software will not take as inputs the raw data, but rather data that has been *preprocessed* to standardize it in some fashion. In the case of image data, this may involve, for example, conversion to standard image resolution, smoothing, reorientation, and intensity normalization. This preprocessing code may need to play a dual role: it may both be part of the data preprocessing (i.e. external to our software) and needed by our actual product software supplied to the user, as the new patient data that will be processed by our software may need the same/similar preprocessing process as the training data. Hence, here we have code that may serve double duty and needs to be managed accordingly. The easiest way to handle this is to place this code in a separate library that is accessed by both the data preprocessing tools and our actual "production" software.

When using supervised learning techniques, one cannot escape the need for *data annotation*. For any operation, the ML model needs to be taught what the correct result is. This is most commonly done by having "expert" human raters annotate the input data to create such ground-truth annotations. This requires training the raters, establishing quality measures for the annotations (e.g. inter-rater variability), and having processes to perform quality assessment of the overall annotation process.

The data acquisition, preprocessing, and annotation steps comprise the activities in step A of Figure 9.3.

At this point we are ready to create a *data set* of matched input data[7] and annotations that can be used to train an actual ML model. As the NMPA states [189], what we are creating here is a training set and a tuning set.[8] The data set may need to be augmented using appropriate data augmentation methods (e.g. image flipping and rotations) to expand the size and to increase the robustness of the training process. This creation of the final data set is shown as step B in Figure 9.3.

9.3.2   Model Training and Initial Evaluation

Once the data set and training code is in place, the next step in the process is to select and train the actual model and perform initial evaluation. The model is trained on the "training" part of the data set (using the "tuning" data to perform algorithm optimization, etc.) The test data is then used for software verification work that ensures that our software works appropriately. This is step C in Figure 9.3. We will discuss validation and its requirements (including the use of separately sourced data sets and third-party evaluators) in Section 15.1.2. This should be performed using our actual complete software package.

As an aside, training most deep learning models requires use of high-end workstations/clusters with powerful graphics processing units (GPUs) to reduce the computational time. Training on GPUs is typically significantly faster (e.g. $100 \times$ faster) than training on standard CPU architectures. Cloud services such as Amazon Web Services (AWS – see Section 9.4) may be used for this purpose if they satisfy the necessary data privacy and security considerations.

In general, the model training code does not constitute part of the software that will ordinarily be supplied to the end user.[9] Therefore, it can be considered part of the development process, that is, we can think of the code used to train the ML model as similar to other software development tools such as compilers (see Section 2.2.2). However, this is code that is used to create outputs that are absolutely critical to the correct functionality of our software. For this reason, such code needs to be carefully controlled, documented, and maintained, and we must have procedures for ensuring that the output is correct.

9.3.3   Deployment and Hardware Constraints

Once the models are trained, they need to be incorporated into our actual software in the same manner as any other module – this is step D in Figure 9.3. Additionally, the preprocessing code may also need to be incorporated in the final software. This will allow the software to take raw data as inputs, as will probably be necessary in a clinical environment (e.g. images directly from the PACS database).

It is worth noting that AI/ML modules often have specific hardware requirements (e.g. GPUs), which may affect/restrict the choice of user environment. In some cases, especially if there are no real-time constraints, the AI/ML modules can be housed on a separate server and accessed by the user over a network interface. This may introduce data privacy and security issues, as discussed in Section 3.4.4.

Our story continues in Section 15.1.2, where we discuss clinical evaluation. Our software now "leaves" the development site and migrates to a testing site (e.g. a hospital). We also need to create a separate evaluation data set (step E of Figure 9.3, which essentially duplicates the process of step A), that is used to independently assess the performance of our AI module (or algorithm). Once this is satisfactory, we can proceed to obtain regulatory clearance and ultimately deploy our software (step G) and monitor it for acceptable performance.

Note: The case of self-improving continuously learning systems algorithms is more complicated [228]. The reader is also referred to the discussion about algorithm change protocols in Section 2.6 and the relevant regulatory guidance documents [82, 234].

9.4 Modern Computing Environments

In this section, we survey the modern computing environment that is available for the software designer in 2021. Fifteen years ago (i.e. around the year 2005), the software developer had three major platforms that they could target: (1) MS Windows – this is dominant in many hospitals and all clinical information technology (IT) would support it; (2) macOS – this is very popular among physicians for their personal computers; and (3) Linux – this is primarily used for servers and scientific computing. Almost all software ran on a dedicated workstation that used one of these three operating systems.

This situation has changed dramatically. In addition to mobile devices (see Section 9.4.3), the software designer has a bigger selection of environments available. The first is the web, which enables the developer to provide applications through a web browser. This has the critical advantage that it requires practically no installation on the part of the user. Modern web browsers are practically complete operating systems in their own right.[10] Probably the first real web-based application[11] was Google's GMail, which debuted in April 2004 and did not become super-popular until a few years after that.

In addition, we have the emergence of cloud-based computing resources such as Amazon's Elastic Cloud or Microsoft's Azure, the use of containerized environments for software delivery (virtual machines, Docker), and the increasing need to support mobile devices such as phones and tablets. This dramatically affects how one designs, implements, and distributes software, and in particular opens the door for many "software-as-a-service (SAAS)" designs that were practically unknown (especially in the clinical environment) not that long ago. We also discuss the implications for the use of such distributed environments for data privacy and security.

9.4.1 Virtual Machines and Containers

When distributing software, one is faced with the painful reality that one's code is only a part (and often a small part) of the software and hardware environment that the user will experience. The environment will consist not only of our code, but also of other components such as the operating system, system libraries, and hardware drivers. Hence, when testing software (especially when the software has critical functionality), one must test the software in all possible hardware/operating system configurations in which one expects it to run. This is a very hard problem. Most medical software vendors solve this problem by supplying the software together with a specific computer that is meant to exclusively run the particular software. In this way, the manufacturer controls the environment completely and ensures that the software environment is fully tested. When this is "delivered," it often comes with many restrictions. For example, the user may be required to not install any additional software or even operating system updates.[12] This solution results in console rooms in many hospitals having multiple computers, each running one piece of software. While this feels "crazy" to most people when they first encounter it, the multiple-computer scenario is simply a consequence of the need to certify that the software environment is valid.

In recent times, a potential solution to this problem has emerged in terms of virtual machines, both of the heavyweight form (e.g. VMWare, Parallels) and of the more modern lighter-weight kind (the primary form of which is Docker containers). This solution typically involves running a piece of software (e.g. VMWare Workstation) on a local machine that, in turn, creates a fully virtualized hardware environment that runs a fully controlled operating system with specific configurations. Then, one distributes one's software as part of a virtual machine that is effectively a completely controlled environment. The software developer is effectively distributing a virtual computer that runs inside a real computer, but is completely isolated from it. While such solutions are less efficient (e.g. one sees 10–20 percent lower computational performance) and require extra memory to run both operating systems, they have become increasingly popular as hardware costs have gone down.[13]

9.4.2 Cloud Solutions

In addition to locally hosted virtual machines, many providers (e.g. Amazon Elastic Cloud, Microsoft Azure, Google Cloud) allow one to move from virtual software machines to literal virtual computers. Effectively, the cloud providers provide a service that allows virtual machines to run on their infrastructure, obviating the need for a powerful "on-premises" physical computer.[14] One can create such virtual machines online and pay (by the minute) for their use as needed and scale up rapidly if there is demand.[15] These services, in addition to supplying low-level virtual computers, also provide higher-level services as preconfigured database servers, application runners, web hosting, data hosting (e.g. Amazon S3), and others.

We think it is important when one is designing a new software solution to be aware of the possibilities of cloud-based delivery and to include this as part of one's design options.

9.4.3 Mobile and Its Challenges

In the past 15 years we have also seen a massive rise in the use of mobile platforms, primarily Google's Android and Apple's iOS (and its variants). This process started with the launch of Apple's iPhone (2007) and subsequently the iPad (2010). These, together with devices running Google's Android OS (mostly phones), now represent the majority of all "computer-like" devices in the world today.[16]

Mobile devices both offer great advantages and present significant challenges to the developer. On the positive side, most people now carry with them at all times a powerful, super-portable computing device. This is true not only in richer countries, but also in low-resource developing countries. For some applications (e.g. screening devices) this opens up the ability to create software that can be used in the field/on location (e.g. in rural settings, at the bedside). In addition, people generally like the simplicity and user-friendliness of the major mobile operating systems, which means that applications that can be made available on these are more likely to be used (and bought!).

From the perspective of those managing clinical IT, mobile devices are also a security nightmare. Most clinical IT/information security offices prefer to keep all data on secured servers housed in special facilities such as data centers. The same applies to workstations that are placed in individual offices. These are centrally managed, encrypted devices, and the users often have no ability to install any additional software on them. The devices belong to the hospital, and the employees (doctor, nurse, support personnel) are simply end users.

With mobile devices, the picture is dramatically different. The devices are typically owned by the users. To quote ISO 90003 [134]:

In some countries employees are encouraged to use personal devices such as mobile phones and computers (bring your own device – byod) rather than those provided by the employer. Employees' own devices can represent a security risk for employers' data and a risk of transfer of malware or computer viruses if poorly managed.

These (from a clinical IT perspective) insecure, uncontrolled devices need to be securely integrated into the clinical IT environment. Users will not accept restrictions on what they install on their own phones, or any other intrusive security settings. Any software that downloads data to a phone/mobile device (e.g. an image-viewing application that connects to a server) must ensure that all data is safe in the case that the device is "lost" or "accessed by unauthorized personnel."[17] One solution is to insist on user authentication (made easier by methods such as face/fingerprint recognition) prior to accessing any data/using the software to ensure no unauthorized use.

A second problem is the issue of software distribution on mobile devices. Both Apple and Google[18] restrict the installation of software ("apps") on such devices to only software that is obtained from authorized sources such as App Stores.[19] Hence, for most software companies, one must obtain approval from Apple and/or Google prior to the installation of any software on a mobile device, which introduces another set of barriers (and potentially risks) that must be planned for and overcome – see the example of cloud gaming services [240].[20]

There are other solutions that rely on the use of special profiles[21] for which companies can apply to allow the installation of private apps from an internal server that is separate from the standard App Store. One such service is Microsoft Intune, which is used by many companies to allow the installation of company-specific apps. The problem with such profiles is that (1) they are intrusive – users need to surrender some control of their devices to their companies; and (2) they are potentially a source of security vulnerabilities.

In summary, mobile devices offer potentially great advantages to the developer, but also some problems, as these devices use operating systems that restrict what a developer can do and what software can be installed on them.

9.5 Software Testing

A reminder: Software testing is not the same as software validation. Software testing is part of software verification – the process of ensuring that software was correctly implemented as designed.

Testing is incredibly valuable. We need to be able to verify the software in addition to validating it. Software testing involves a number of different components, including both automated regression testing and interactive user testing. Regardless of how comprehensive a set of tests is, it can never hope to catch every single error in the system.

A detailed description of these testing techniques can be found in any standard software engineering book and will not be covered here. The key in a regulated software development setup, such as the case of medical software, is that all testing procedures are documented and the reports of running the complete set (especially the interactive user testing) are archived for future reference and review.

There are two basic ways to perform software testing. The first is to run the software and see that it produces the results (computational outputs, visualizations, etc.) that it is supposed to. The second is to run the software (ideally in an automated fashion) in many different ways (i.e. by varying the inputs) and to ensure that the outputs produced are the same as those produced by the software last time the test was run. This second step is known as regression testing. The first type of test ensures that the output of the software is correct, whereas the second ensures that changes to the software did not break the software in any meaningful way. Changes may also involve testing the software in a different operating system environment and/or hardware. Naturally, there is significant overlap between these two types of testing – we always hope that our previous results were in fact correct and not just the same!

A second distinction is between so-called "black-box" and "white-box" testing. In black-box testing, the person performing the tests has no knowledge of the internal structure of the software. In this scenario, what is being tested is the "whole thing," end-to-end. In white-box testing, by contrast, we use knowledge of the structure of the software and, hence, we can design tests for specific functionality. Both are useful and important techniques.

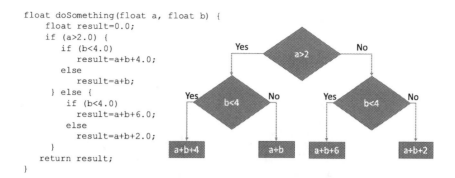

```
float doSomething(float a, float b) {
    float result=0.0;
    if (a>2.0) {
        if (b<4.0)
            result=a+b+4.0;
        else
            result=a+b;
    } else {
        if (b<4.0)
            result=a+b+6.0;
        else
            result=a+b+2.0;
    }
    return result;
}
```

Figure 9.4 The multiple paths problem. The code (left) and the equivalent flowchart (right) illustrate the multiple paths problem. This is a key confound in testing. Our code operates in four different regimes depending on the inputs. Proper testing, therefore, requires testing with inputs from each of these regimes.

9.5.1   The Multiple Paths Problem

Consider the pseudo C code and flowchart shown in Figure 9.4. This rather trivial piece of code presents a significant challenge from a testing perspective. It has two nested "if" statements which result in our simple function having four different modes of operation. Hence, to test this code completely, we need to specify at least four input conditions ($a > 2$ and $b < 4, a > 2$ and $b \geq 4, a \leq 2$ and $b < 2, a \leq 2$ and $b \geq 4$). As one can easily imagine, the situation gets significantly more complicated and challenging with more realistic code. The GPSV [70] makes exactly this point when it states:

Unlike some hardware failures, software failures occur without advanced warning. This is because software can follow many paths (due to branching statements) some of which may not have been fully tested during validation.

There are two solutions to this problem. One must either have a very detailed testing regimen to capture all possible combinations, or prune some paths. Branches often arise from the software having "extra" functionality. In mission-critical cases, the usual solution is to eliminate complexity and restrict the software to a smaller feature set that makes it easier to test. As one becomes more experienced in software development, one becomes more aware that one should write code that has the minimum amount of features. Such code is more likely to be conducive to comprehensive testing.

9.5.2   What Is Software Testing?

A good definition of software testing is given by Vogel [261, p. 256]:

An activity in which a system, subsystem, or individual units of software is executed under **specified conditions**. The **expected results** are anticipated from requirements or designs. The

actual results are observed and recorded. An **assessment** is made as to whether the actual results are acceptably equivalent to the anticipated results.

To expand: When we create a test, we first have a set of *specified conditions*. These specify the inputs of our software.[22] In the trivial example from the previous section (see Figure 9.4) these are the values of a and b.

Next, we have the expected results. We should know what the result is going to be given a and b based on the design of the software. In more complicated cases, one tests the software manually many times until the result "feels good" (e.g. in the case of visualizations) and then specifies this "good output" as the expected result – this procedure is often the basis of regression testing.

Finally, we run the software and record the actual results. These results are then compared ("the assessment part") to the expected result. If the results are close enough[23] to the expected results, we declare the test a success, otherwise we declare that it has failed. *This is the basis of all software testing.*

9.5.3 Levels of Testing

The paradigm described above can be applied at all levels of software testing. Following IEC 62304 [119], we define three levels of testing:

1. **Unit testing:** This is where individual pieces of code are tested independently for correctness. For example, our code may have a function that computes the mean of a set of numbers. Our test will call this function with known inputs and compare its output to the known output. This is clearly white-box testing and involves writing additional code (the testing code) that will exercise the software.

2. **Integration testing:** This is where we focus our testing on how well data and control are transferred across the software's internal and external interfaces. This may involve testing correctness of database access (where the database is an external system) or the communication between the user interface and the computational code. Testing at this level often involves writing simulators (e.g. a function that mimics the GUI and provides input to the computational code) to see if the internal subsystem is functioning correctly. Again, this is white-box testing, as we need to know what the interfaces are.

3. **System-level testing:** Here we test that the software as a whole works in an environment that is as similar as possible to the actual user environment. This is black-box testing, as the tester does not have any knowledge of the internals and ideally is completely independent of the software team. System-level testing benefits also from a degree of randomness, i.e. the testers should be encouraged to also perform some random testing to see if they find any additional issues.[24]

We will have more to say about testing in Chapter 14 in the context of software verification.

9.6 Source-Code Management

The best practices for managing source code involve some form of revision control system. Revision control systems allow users to maintain multiple versions of the code, manage revisions, backtrack to previous versions, and collaborate with others in a team. For those not familiar with the concept, they are a form of shared folder system (e.g. Dropbox) optimized for sharing code. Revision control systems provide the user with the ability to revert to any previous version of a source code file, effectively undoing any changes performed at any point.

The original revision control system was SCCS, which originated in 1972. This was succeeded for most people by RCS (1982), which worked on single files – that is, each file had its own version number and history. This was still in use all the way into the late 1990s. CVS (Concurrent Version System) began its life in the late 1980s and was the first whole-project system running on a server. This was a great advance on what had gone before, as it allowed centralized version management. Subversion came along in 2000 and was probably the most popular system until about 2016; it fixed many issues with CVS, including having a single easy version number for the entire repository.

At the time of this writing (2020), most popular systems are server-based, which is critical for collaboration. By far the most popular server system is GitHub (www.github.com), which uses the Git revision control software.

9.6.1 Git and GitHub

We present here a very brief introduction to the Git revision control system (and by extension GitHub). There is much excellent material on Git online (a quick Google search will point you to this, but see also Chacon and Straub [29]), so our goal here is mostly to introduce the concept to someone unfamiliar with the whole concept of revision control and provide them with some background material. All of this explanation assumes that one is using a Unix shell (bash) command line. This can be accessed using the terminal program on macOS or the Git-bash console on MS Windows (this is part of the Git distribution).[25]

One way to look at Git is as "Dropbox on steroids." It allows for multiple users to share files (primarily but not exclusively code) and revisions. The official[26] copy of the code lives on the server in what is known as a repository (think of this as the shared Dropbox folder). Assuming that the repository already exists, the process of using Git goes as follows (see also Figure 9.5):

- You, the programmer, create a linked copy or a "clone" of the repository on your local machine. This is a full copy of the repository.
- You work on your computer, make changes, and save or "commit" these changes to your local copy (clone) of the repository.

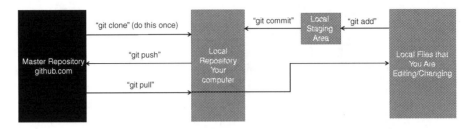

Figure 9.5   The Git/GitHub workflow. This is main workflow when using Git and GitHub. Note the four locations in which files can lie. From left to right, we have the server, the local repository (complete copy of the server), the staging area, and the actual files that the user is editing.

- You can synchronize your changes back to the repository by "pushing" your changes to it.
- You can get the changes that others made to the repository by "pulling" the latest updates down to your local clone.

The items in quotes (e.g. "clone") are in fact the Git sub-commands that one uses to perform these operations. For example, to clone a repository in a local folder (which we will call "FOLDER"), one would execute the following commands:

```
cd FOLDER
git clone URL_OF_REPO FOLDER
```

Next, consider the case of a file called "somecode.cpp" in the repository that one edits to make a small change. When the work is done (or at least progress is made), one first adds the file to the local staging area using:

```
git add somecode.cpp
```

Then, the change in the code is committed to the local repository as follows:

```
git commit -m 'change to code to enable new feature'
```

This cycle (git add, git commit) may be repeated multiple times. When the code is done, we can push the change to the repository:

```
git push
```

Periodically, one then also updates one's local version of the code from the main repository:

```
git pull
```

There are some issues when multiple people make changes to the same file(s) simultaneously, introducing conflicts. This is an advanced topic – simply search for "git conflict resolution" online for more information. In addition there are many excellent "cheatsheets" for git available online – just Google "git cheatsheet."

9.6.2 GitHub as a Development Platform

Branches and Commenting

A more advanced workflow than the basic setup described above involves the use of branches. Here, when we need to make significant changes to the code, we follow the following procedure (see also Figure 9.6):

1. We create a branch – a separate version of the code that lives in the same repository but is separate from the core repository ("the main branch" – this used to be referred to as the "master" branch).
2. We switch our editing to the branch.
3. We perform our new work, make changes, commit, push, etc.
4. When the work is completed, one creates a "pull request" on GitHub to request that the changes made in the branch are merged back into the main branch.
5. At this point (in a team environment) another, potentially more senior, developer reviews the changes and highlights issues, adds comments, and so on. This may involve multiple people.
6. The original developer (or somebody else, perhaps) responds to the comments and makes the necessary changes.
7. Once this process runs its course, the changes are approved and the branch code is merged into the main repository.
8. The branch is then deleted – it is always meant to be transient.

This is an example of a process that, in the beginning, feels "over-the-top." Why should one need to do all this for what often is a minor change to the code? The reason (especially in a medical software context) is that this, in addition to being an excellent process to follow in general, also has the massive advantage of being completely compatible with a QMS (see Chapter 4).[27] All changes are documented, including who did the work, who reviewed it, and who approved it. This effectively implements the standard "review, revise, and approve" (RRA) process. The process is fully documented in the repository and can be re-examined and revisited in the future.

Regression Testing via GitHub Actions

An additional feature of GitHub is the use of GitHub Actions to perform nightly regression testing. An action is a job that can be run by GitHub in a virtual environment, either on schedule or triggered by some event (e.g. on push). An action is defined by creating a YAML file. To create an action, one begins by creating a GitHub workflow file in the `.github/workflows` subdirectory of a repository. Here is an example from the bioimagesuite web repository that invokes a Docker job:

```
on:
    schedule:
      - cron: '13 6 * * *'
jobs:
  docker_test:
    runs-on: ubuntu-18.04
    name: Docker test for bisweb
```

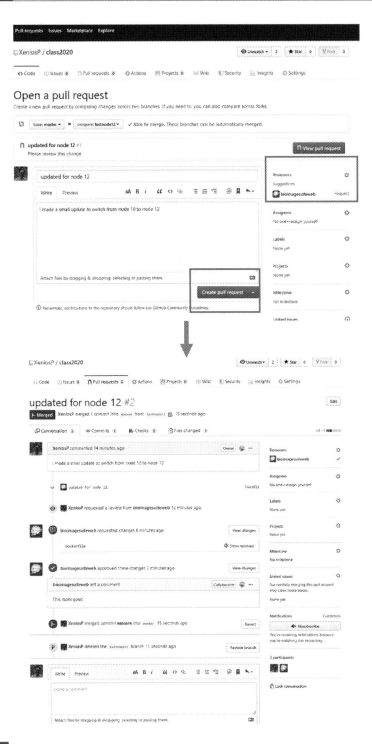

Figure 9.6 GitHub review workflow. The first and last steps in the GitHub branch–review–merge workflow. The picture on the top shows the initial request for the review. The picture on the bottom shows the log for the overall process, detailing who said what and who performed the final approval.

```
steps:
- name: Run
  id: run
  uses: bioimagesuiteweb/bisweb/actions/docker@devel
- name: Results
  run: echo "${{ steps.run.outputs.result}}"
```

The Docker job is defined in a separate file (../actions/docker/action.yml in this case), which reads as follows:

```
name: 'Hosted YML'
description: 'Regression test'
outputs:
    result: # id of output
      description: 'The abbreviated results file'
    logfile: # id of output
      description: 'The complete log file'
runs:
  using: 'docker'
  image: './Dockerfile'
```

The job is defined in a dockerfile that is placed in the same directory as action.yml that can be used to define the environment and run the job as needed.

The action need not use Docker. GitHub supports hosted Linux, MS Windows, and macOS environments that can be used to create and run jobs for a variety of needs. This is a lesser-known part of GitHub that could have a significant impact on a software project, as it can obviate the need to have local machines dedicated to running regression testing, with all that this entails.

9.7 Conclusions

A medical software project of any reasonable size cannot be developed by simply having a single developer "hack away" until he or she has something that does the job. It requires the coordination of many people using a structured process that is organized around the use of a life cycle model. The use of AI/ML modules extends the need for organization to the management of large amounts of data (as opposed to just code) and the need to have processes in place for the creation of training data and its use to train ML models. Similarly, software testing and code management require planning and the use of appropriate techniques. While software life cycles and software testing are explicitly mentioned in regulatory documents, source code management comes under the purview of an organization's QMS.

9.8 Summary

- The software life cycle is the organizing principle of any large software project. There are many different software life cycles, such as the traditional Waterfall model and the more modern Agile model.

- When the software includes the use of modules based on AI/ML supervised learning techniques, the software engineer must account for these in the life cycle process. A critical additional part of the process is training data management.
- Modern software can be run on a variety of platforms, including virtual machines and containers, cloud-based solutions, and mobile devices. Each of these platforms has its own benefits and challenges that need to be addressed.
- Software testing is an integral part of the software development process. We distinguish between white-box testing (performed by the developer with full knowledge of the code) and black-box testing (performed by a tester with no access to the code). Both are important in identifying errors and other problems with our software.
- Finally, source-code management is a critical component of any software project. The most popular tool for doing this at the time of writing is Git in combination with the GitHub service. Such tools can be used not only to manage code, but also as part of an organization's quality process as they allow for reviews and approvals of code changes.

RECOMMENDED READING AND RESOURCES

An excellent brief introduction to software life cycles can be found in a brief online article by Alspaugh:

T. A. Alspaugh. Software process models.
www.thomasalspaugh.org/pub/fnd/softwareProcess.html.

For more advanced readers, the two core industry standards in this area are:

IEC. *IEC 62304: Medical device software – software life cycle processes,* 2006.

ANSI/AAMI/IEC 62304:2006/A1:2016 Medical device software – software life cyle processes. (Amendment), 2016.

Association for the Advancement of Medical Instrumentation (AAMI). AAMI TIR45:2012/(R)2018 Guidance on the use of AGILE practices in the development of medical device software. Washington, DC, August 20, 2012.

Agile techniques are becoming increasingly popular, but also have their own problems, as discussed in the following book:

B. Meyer. *Agile! The Good, the Hype and the Ugly.* Springer, 2014.

There is an exploding interest in the use of AI/ML in medical software. This is a rapidly evolving area. A good regulatory document is the German National Standard:

DIN SPEC 92001-1:2019-4 Artificial intelligence – life cycle processes and quality requirements – part 1: quality meta model April 2019.

This Microsoft case study on software engineering and AI describes work at one of the largest software companies in the world in this area:

S. Amershi, A. Begel, C. Bird, et al. Software engineering for machine learning: a case study. In *Proceedings of the 41st International Conference on Software Engineering: Software Engineering in Practice,* ICSE-SEIP '19, pp. 291–300. IEEE Press, 2019.

A good review of the issues related to continuously learning systems can be found in the following white paper:

B. Sahiner, B. Friedman, C. Linville, et al. Perspectives and good practices for AI and continuously learning systems in healthcare, 2018.

There is ongoing movement toward explainable AI. This is an attempt to overcome the "black-box" nature of most modern ML algorithms. Two good sources are:

E. Tjoa and C. Guan. A survey on explainable artificial intelligence (XAI): towards medical XAI. arXiv:1907.07374, 2019. http://arxiv.org/abs/1907.07374.

C. Rudin. Stop explaining black box machine learning models for high stakes decisions and use interpretable models instead. *Nat Machine Intell*, 1(5):206–215, 2019.

There is also some fascinating work whose goal is to use AI/ML to learn the system requirements directly:

A. Vogelsang and M. Borg. Requirements engineering for machine learning: perspectives from data scientists. In *2019 IEEE 27th International Requirements Engineering Conference Workshops (REW)*, pp. 245–251, 2019.

The book by David A. Vogel is an excellent discussion of software testing focused on medical applications:

D.A. Vogel. *Medical Device Software Verification, Validation, and Compliance*. Artech House, 2011.

Finally, for those students new to revision control, the following book provides an excellent introduction to the topic:

S. Chacon and B. Straub. *Pro Git*, 2nd ed. Apress, 2014.

NOTES

1. Some still have the false impression that most software is created by somebody simply siting at a computer and "hacking away" until they have a solution. This is still unfortunately true for some software projects. One should be wary of using software that was created in this way. As Alspaugh [7] points out, the simplest model that is used (often implicitly) by "unsophisticated" developers is "code-and-fix." This cannot be recommended for any large project where we need quality management and assurance.
2. One of the benefits (or consequences) of having multiple development cycles in the last two models is that we have usable interim software builds. These may be made available to the users both for testing and for the evaluation of how well the software meets their needs.
3. Working prototypes are important because they enable users to provide feedback. Marketing departments can also use working (but not fully finished) prototypes to create demand for the final product.
4. If interpreted literally they are, in fact, incompatible with regulatory guidance.
5. Algorithm researchers face similar problems. Historically, algorithm developers could focus on mathematical formulations and algorithm implementations only with large

data sets coming in at the end of the process. The use of ML techniques has upended this style of work.

6. This is also sometimes referred to as the "truthing" process.

7. The input data are the outputs of the preprocessing of the original acquired data.

8. In the ML literature, we often read of the data set being divided into three parts: (1) a training set used to train the model, (2) a validation set used to tune the model's parameters, and (3) a test set used at the end to evaluate the final model. As the NMPA states [189], the word validation means something completely different in the medical software domain, so they (per the unofficial translation) propose the word "tuning" instead, which we adopt here.

9. An exception to this rule is the case in which our software will have the capacity for self-improvement, which introduces additional regulatory issues such as algorithm change protocols (ACP) – see Section 2.6.

10. One of them, Google Chrome, in fact did morph into an operating system: ChromeOS, which is very popular in educational settings. It primarily ships on laptops known as Chromebooks. These originally did not permit the installation of any additional software other than web-based applications. Note, however, that some web-based applications can be considered SaMD and are regulated as such [234].

11. By "real," we mean here an application where a significant amount of the processing is done client-side (i.e. on the user's computer) in the browser in JavaScript as opposed to the older style of web applications which basically had to invoke the server for every little processing step.

12. This is the reason why one finds computers in hospitals running ancient versions of Windows which, additionally, are often missing critical security updates. This is effectively a case of the vendor passing the problem on to the customer.

13. In 1995, at our lab at Yale, a "high-end" machine had 1–4 GB of disk space and 4–64 MB of RAM. Now a machine that costs significantly less can have multiple terabytes of disk space and 32–128 GB of RAM.

14. One still needs a computer to access the remote service, but such computers often have minimal hardware requirements.

15. Netflix (as of 2020) was one of the biggest Amazon Cloud customers.

16. Smartphones are effectively pocket computers.

17. A common example is children of employees borrowing a phone to play games!

18. Between them, these two companies control the operating systems running on the overwhelming majority of all mobile devices.

19. Google Android permits the direct installation of applications if one turns certain security features off. This is not recommended, however.

20. See also the case of the Iowa Caucus app described in the vignette presented in Chapter 20. Here, the developers abused the beta-testing distribution mechanisms provided by both Apple and Google to bypass the normal distribution process [36].

21. These are essentially authorizations provided by Apple to specific companies, that permit lower-level access to the operating system, including (in some cases) remote

access to the data on the device and the installation of applications from private application stores.

22. Software is assumed to be deterministic. The outputs should be completely specified by the program (which is fixed) and the inputs. Inputs, naturally, means more than simply the actual variable values, and could include our environmental parameters, such as CPU load if the test is trying to measure computational performance.

23. The definition of "close enough" is application-specific.

24. We find that good procedures for system-level testing are often similar to those used for prostate cancer biopsy. Currently, the procedure for prostate biopsy consists of two steps. In the first step, the urologist targets known lesions visible in preoperative images. Next, he samples randomly from different parts of the prostate. Sometimes, these second biopsy samples are the ones where significant cancer is found.

25. A useful, mostly unknown, detail is that one can open a Windows folder or a macOS folder from the command line using `start .` on MS Windows or `open .` on macOS.

26. The term "official" reflects the typical policy in most companies as opposed to any restriction placed by Git itself, where all repository copies are equally valid.

27. In fact, the procedures for committing to a repository and merging branches should be explicitly defined in the QMS documents of the organization.

Part III

An Example Medical Software Life Cycle Process

10 The Overall Process

INTRODUCTION

This chapter presents an example medical software life cycle process. We first introduce the topic (Section 10.1), and introduce our example project – the image-guided neuro-navigation system – that we will use to illustrate the process over the next few chapters. Next, we summarize our life cycle "recipe" (Section 10.3) for medical software. This recipe follows the guidelines in the international standard IEC 62304. It begins with identifying user needs and concludes with the satisfaction of these needs via a structured process of first establishing requirements and then designing, implementing, testing, and releasing our software. We also briefly discuss software safety classifications (Section 10.4) and present some thoughts on traceability and planning for the overall process (Section 10.5). The chapter concludes with a section that uses the children's game "Telephone" to illustrate some of the potential pitfalls in this long process (Section 10.5.4).

10.1 Overview of the Software Life Cycle Process

In this third part of the book, we present a "recipe" for designing a medical software package that is appropriate for an undergraduate class. This is by necessity a simplified version of the real process and should not be used as an explicit template by a new company in designing a product. It is meant rather to illustrate the principles covered in earlier parts of the book and to provide a concrete example.

The "product" we will design is a software for image-guided neurosurgery (IGNS) (Figure 10.1). This is not meant as a new or innovative design. This is an established product category, and such products have been in use for at least 20 years, beginning in major academic hospitals and progressively expanding to smaller medical centers. The most popular systems in this area are the Medtronic StealthStation [175] and the BrainLAB VectorVision Cranial system [22]. Anyone familiar with these will soon realize that what we are "designing" here is a simplified version of these products. One of the earliest descriptions of image-guided neurosurgery was by Grimson et al. [103].

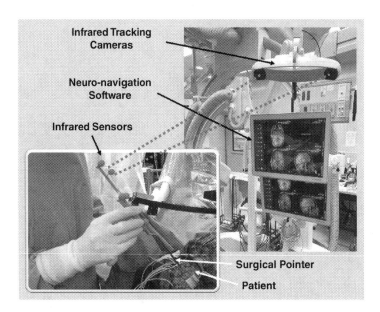

Figure 10.1   An image-guided neurosurgery system. This picture shows the key components of an image-guided neurosurgery setup. At the top-right of the figure is the neuro-navigation software – this is what we will be trying to design. The goal of such a navigation system is to perform surgery of the brain in a safer and more accurate fashion, providing the surgeon with real-time feedback as they operate. The software receives input from the infrared tracking cameras (top) which track, in real-time, the position of the surgical tool (shown zoomed in, bottom-left) and displays its position relative to the preoperatively acquired MR images of the patient on the screen. This is similar to how GPS software shows the position of a car on a city map to allow a driver to navigate. The picture from the operating room is courtesy of Dr. Dennis Spencer, Yale School of Medicine.

10.2   The IEC 62304 Standard

The IEC 62304 standard, originally issued in 2006 [119] and revised in 2016 [2], is the primary document that describes how medical software should be developed. It is accepted as the recognized consensus standard by the FDA [83] and other regulatory agencies.

The goal of the standard is to provide a "framework of life cycle processes with activities and tasks necessary for the safe design and maintenance of medical device software." It then states explicitly that it assumes that the software is developed and maintained within a quality management system (see Chapter 4) and a risk management system (see Chapter 5). The point is clear. One cannot develop safe medical software without a quality and a risk management system. They are absolute prerequisites. A second example is the statement that "medical device management standards such as ISO 13485 ... and ISO 14971 ... provide a management environment that lays a foundation for an organization to develop products."

The original edition of the standard consists of nine sections:

1. Scope (1 page);
2. References (1 page);

3. Terms and Definitions (4 pages);

4. General Requirements (2 pages);

    4.1 Quality Management System;

    4.2 Risk Management;

    4.3 Software Safety Classification;

5. Software Development Process (10 pages);

6. Software Maintenance Process (2 pages);

7. Software Risk Management Process (1 page);

8. Software Configuration Management Process (1 page); and

9. Software Problem Resolution Process (2 pages).

In addition there are four annexes (Annex B is particularly valuable):

A. Rationale for the Requirements of the Standard (3 pages);

B. Guidance on the Provisions of the Standard (16 pages);

C. Relationship to Other Standards (21 pages); and

D. Implementation (2 pages).

The heart of this document, the description of the overall software development process, is summarized in Figure 10.2. This figure, which is adapted from the IEC 62304 [119], illustrates the complete process from "user needs identified" (top-left) to "user needs satisfied" (top-right). In Section 10.3 we will discuss each of the elements involved in this process, and in Section 10.4 we will focus on the critical topic of software safety classification.

We note that one problem with the current iteration of the standard is the absence of guidance on the use of artificial intelligence/machine learning (AI/ML) methods and how these affect the overall life cycle. We provide a discussion of this topic in Section 10.5.3. Hopefully, future versions of this standard will address these issues in some detail.

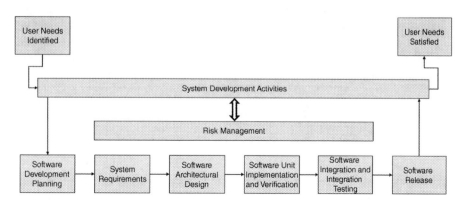

Figure 10.2 The Software Process According to IEC 62304. This figure illustrates the software development process. (This figure is a simplified version of a flowchart that appears in IEC 62304 [119].)

10.3 The Life Cycle Process

Before we present an example recipe for a life cycle process, we need to address a common misunderstanding. Most regulatory guidance and the IEC 62304 standard appear to suggest the use of a Waterfall life cycle model – this is described in Section 9.2 and in particular Figure 9.1. The choice of this life cycle (both in the standards and in this text) is primarily driven by its simplicity from the perspective of actually explaining it, as opposed to any claims that it is superior to other approaches. As IEC 62304 states explicitly (Annex B [119]):

This standard does not require a particular SOFTWARE DEVELOPMENT LIFE CYCLE MODEL. . . . Because of such logical dependencies between processes, it is easiest to describe the processes in this standard in a sequence, implying a "Waterfall" or "once-through" life cycle model.

Agile techniques are also perfectly appropriate (and some would argue better [143, 212]) for the development of medical software, as discussed in the AAMI technical report TIR 45 [1]. These methods are being widely used.[1] We discuss Agile methods in Section 9.2.2. In any event, the principles described here are equally applicable to all life cycles. What follows is a simplified recipe created for this introductory textbook.

10.3.1 User Needs

First, in Chapter 11, we will discuss the needs-finding process. What do our users actually need? Imagine yourself as an engineer walking into a neurosurgery operating room in the year 1995 (the pre-IGNS era) and trying to imagine what a product should look like that will address your client's (i.e. your favorite neurosurgeon's) needs. What are they trying to do right now? How are they doing it? What is it that they would like to be able to do? How will they be able to do it using your product?[2]

Establishing user needs is not a regulated process. This is where people brainstorm, discuss ideas, dream, and try to visualize what the "brighter future" will look like. At this point, one may well need to go back to the "lab," build some prototypes, play with new technologies, and in general identify what is possible. This might be the research component of the so-called R&D (research and development) process. For the purpose of this book, we will not discuss what this research might entail.

10.3.2 System Requirements

Next begins the formal process. We first need to establish the requirements for our system, as discussed in Chapter 12. What are those "things" that the system **shall** do? Critical to this is the identification of the usage scenarios for our system. In teaching, we have used the game Clue (see Figure 10.3) as an analogy. A usage scenario can be summarized as the answers to the following questions:

Figure 10.3 The board game Clue (or Cluedo). In this game, the players try to solve a crime and, in particular, to identify **who** did the crime, **where**, and **with what** weapon. This, in our mind, is a great analogy for establishing the use cases for a piece of software. These can be summarized as "who" needs to do "what" (hopefully not a crime!), "where," and "with what."

- **Who** is our user? (e.g. doctor, nurse, patient);
- **What** do they need to do? (e.g. look at images, perform surgery);
- **Where** will they be doing it ? (e.g. office, operating room, home);
- **With what** device will they be doing it? (e.g. desktop computer, touchscreen PC, phone).

Once these scenarios have been identified, we can create the set of requirements for our system. These are the capabilities of our system that are needed in order to satisfy the users' needs. If the user, for example, needs to look at medical images, clearly one of our requirements is that our system shall be able to (appropriately) display medical images. Some user needs map to more than one requirement and some requirements satisfy more than one user need.

In this and all other documents one must resist the urge to treat them as acts of creative writing. It is best to use simple language and avoid the use of synonyms to refer to the same operation. "Test" should be "test" everywhere. Using synonyms such as "check" and "confirm," will only add confusion, as the reader (who needs to use this as an input to their own design) may end up trying to understand differences that were not meant to be there in the first place.[3] It is also important to keep this type of document relatively short so that it will be read as opposed to just skimmed, but not so short that it is not sufficiently complete.

One final note is appropriate here: All specification documents may need to evolve as the project evolves, since the actual implementation process may reveal issues that were not identified in an earlier phase. However, specifications must evolve in a controlled manner to satisfy regulatory requirements. Such documents should therefore have a history section that lists when it was first written (and by whom), when it was approved, and when (and why) it was revised. They should always reflect the current state/design of the project. If, during development, one identifies an issue with the specifications, the proper procedure is to stop the development, revise, review, and formally approve the changes to the specifications and then proceed with the development work. This is the way to ensure that we are operating "under control," consistent with regulatory guidance.

The system requirements document is the key to the whole project. Once this is completed, it can be used to create a roadmap for all further processes related to the software. From here, we can begin the process of detailed software design (the creation of the software design document – see Chapter 13). This will provide a more detailed design of our software, including sketches of the user interface and step-by-step workflows for all the processes.

10.3.3   Software Design Document

We discuss the software document in Chapter 13. The software design document provides the roadmap for the actual implementation of the software. The system specifications can be thought of as the bridge between management and the senior software designers. Similarly, the software design serves as the bridge between the senior software designer(s) and more junior software engineers that will be tasked with the actual software implementation (i.e. the writing of actual code).

A possible structure for this document is as follows: First, we present an overview/summary of the software and potentially even a summary of the significance of the project. Then we list the inputs and the outputs of the software and identify (if appropriate) its place in the larger clinical workflow. Next, we present the "big picture design," which should include an overall workflow or use case flowchart (or perhaps multiple flowcharts) and a linkage of these workflows to system requirements. Again, for completeness all workflows must map to system requirements and all system requirements must map to software workflows. In addition to the flowcharts, it may be appropriate to provide an outline sketch of the user interface especially if this is critical to a particular user need.

Following the big picture, one then proceeds to describe an item-by-item specification of the workflows. This description should include some background information as to what we are doing, describe the actual procedure, and then describe how this should be tested and what could go wrong. This will allow for the testing verification of the software. The final sections of this document should describe any external dependencies of the software (e.g. libraries). It should also state how the software will be packaged, labeled, and distributed, and items pertaining to any other regulatory requirements.

In general, a good software design document makes the actual coding task easy. One can simply proceed to implement one module at a time and then generate tests to demonstrate that the module is correctly implemented and integrates into the larger context. Close adherence to the design (which can be enforced by proper reviews) can be used to ensure the avoidance of functionality creep (i.e. the addition of extra non-required functionality).

Please note that the overall process is independent of the actual programming language/programming environment used. These are common principles and apply equally well to projects implemented in languages from C++ to JavaScript.

10.3.4 Construction and Verification

In Chapter 14, we discuss software construction, testing, and verification. Construction (or coding) is the process of actually implementing the design. Verification is defined as ensuring that the software is correctly implemented as designed. This is to be contrasted with validation, which is defined as ensuring that the correct software was implemented and that it meets the needs of the user. The two are often, mistakenly, used interchangeably.

10.3.5 Validation

In Chapter 15, we discuss the validation process. Validation is the process by which we ensure that the correct software package was developed. The validation specification takes the (hopefully quantitative) evaluation criteria from the requirements in the system specifications document and proceeds to describe a process through which one can demonstrate that these are met.

One way to write a validation document is to structure each test as a formal scientific experiment. This begins with an explicit statement of the hypothesis of the experiment, such as: "To demonstrate that the new technique significantly outperforms an existing technique by improving accuracy by 30 percent." Then we proceed to discuss the rationale for our experiment/test. This provides some background material and explains how (in this case) 30 percent improvement was chosen as the appropriate metric.

Following this we proceed to describe the experimental procedure to be followed. This should end in a set of measurements. A figure and/or flowchart might help here. Next, we take the set of measurements and analyze them in some fashion (possibly involving some statistical techniques) to prove that we have successfully established our hypothesis.[4] The final part of this description is a description of alternative strategies that could be followed if the proposed procedure proves impossible to execute. This may be omitted if this is a standard procedure with which we have extensive experience though it might still be best to state that explicitly.

10.3.6 Deployment, Maintenance, and Retirement

Finally, in Chapter 16, we briefly discuss the final three steps: (1) deployment, (2) maintenance, and (3) retirement. The life of a piece of software does not end once it leaves the software engineering environment. Just like for a teenager leaving home, this is when its "real" life begins.

First, one has to create processes/plans for the actual deployment of the software. How does the software gets into the hands of the user? How does it get installed and integrated in the user's environment? What is the training process for new users?

Once the software is in place, it must be maintained. A key component of this is updates, which are a major hazard, as updates can introduce new behaviors and bugs.

The other aspect is monitoring the use of the software to ensure it is behaving correctly, both to enable correction of faults and as an input to improvement processes.

Finally, the time comes for the software to be retired. At some point in the life of the software, we stop maintaining it and move on. Updates are no longer provided. This is also an important step that requires procedures for ensuring patient safety during this phase and the safeguarding of patient data and any other confidential or valuable data created during its use. The key to this process is communication with sufficient warning to enable users to transition to newer solutions.

10.3.7 Risk Management

Embedded in this process, at each stage, is risk management. At each step, one must identify hazards and plan for how to eliminate or mitigate the risks they cause. This should include an evaluation of cybersecurity issues that may arise – see Wirth [269] for more details. The process must, naturally, also be documented appropriately.

10.4 Software Safety Classifications

An important aspect of how the software will be developed is the software safety classification.[5] IEC 62304 [119], as amended in 2016 [2], divides software into three classes[6] as follows:[7]

- **Class A:** The software system either cannot contribute to a hazardous situation, or, if it can, that situation does not result in unacceptable risk.
- **Class B:** The software can contribute to a hazardous situation which can lead to unacceptable risk, but the resulting harm cannot cause serious injury.
- **Class C:** Otherwise.

IEC 62304 (A1) has a very nice decision tree/flowchart (figure 3 in that document) that guides the software "maker" in the classification process. Essentially, the software always starts out as Class C. Then the first question is: "Can a hazardous situation arise from failure of the software?" [2] If the answer is "no," then we have Class A. The second question has to do with whether the risk is unacceptable (after control measures). If the answer is also "no," we are also in Class A. If the risk is unacceptable, then the final classification has to do with the severity of the possible injury. If this is non-serious, we end up in Class B, otherwise class C.

This classification is critical, as the requirements for the software process escalate from Class A to Class B and Class C. For example, in Section 5.4 of IEC 62304, which is titled "Software detailed design," we have a subsection (Section 5.4.1: Subdivide software into SOFTWARE UNITS) which states:

The MANUFACTURER shall subdivide the software until it is represented by SOFTWARE UNITS. [Class B, C].

Please note that this requirement is only for Classes B and C. Class A software does not require this level of detail in the design. This type of statement can be found throughout the standard. It clearly signifies that the process must be more rigorous for software that has the potential to cause more serious harm. This is simply common sense.[8]

10.5 Other Issues

10.5.1 Traceability

Traceability is an important requirement of the regulatory process. Traceability is defined in IEC 62304 as the "degree to which a relationship can be established between two or more products of the development process" [119]. A traceability analysis creates (traces) a map of relationships from user needs all the way to risk verification and validation, and also risk analysis. It also ensures that all work done (e.g. code, testing) links back to a user need and forward to a verification and validation procedure. A search for the term "traceability" in the GPSV (see also Vogel [261], who has a nice map of the interconnections) reveals a number of connections. A partial list is:

1. System requirements[9] to risk analysis.
2. System requirements to software design:

   - The software design implements all the requirements. This ensures completeness.
   - All aspects of the design should trace back to the requirements. If this is not the case, it may reveal that the requirements do not sufficiently describe our software, or that we are implementing unnecessary features beyond what is needed by the requirements.

3. Source code traceability:

   - Our software design is completely implemented in code.
   - The implemented code trace backs to the software design and forward to risk analysis.
   - Our code traces forward to appropriate tests.

4. Testing traceability:

   - Unit tests trace back to detailed software design.
   - Integration tests trace back to high-level software design.
   - System-level tests trace to system requirements.

An additional factor is that one should be able to ensure traceability of the final software product to the actual source and processes used to create it [234], to enable appropriate correction/review measures should the need arise.

A traceability analysis can also help identify what might be affected if the requirements need to be changed [261]. This is particularly important when the software must be changed/updated for any reason. Traceability is often performed using dedicated software tools. It is a required component of the regulatory submission process – see Section 2.3

10.5.2   Planning

One of the important aspects of having an explicit life cycle model in mind is that one is then reminded to think and plan about the entire life cycle prior to actually beginning the detailed work of the project. Working backwards: If our product cannot be deployed,[10] then proceeding with the project is pointless. The same applies to the situation in which our proposed software cannot be validated because we lack, for example, facilities, collaborators, and access to hospitals. It is important, therefore, to first work through the life cycle of the project and try to identify any key problems before investing the time, money, and effort necessary, only to discover that there is a "show-stopper" at the end.

10.5.3   Issues Related to the Use of AI/ML Techniques

If we plan to rely on machine/deep learning techniques as part of our software, we must be cognizant at the beginning of the process of both the training data requirements and any hardware constraints that the use of these may introduce to our project. These topics are discussed in more detail in Section 9.3.

In particular, the data component of AI/ML may also complicate the traditional division between system requirement and software design. In general at the systems level we focus on abstract needs, such as "quantify tumor volume from images." However, if AI/ML methods are involved the availability of training and validation data (and collaborating institutions for validation – see Section 15.1.2) and their limitations may impose constraints on the rest of the process (such as the software design). For example, if we only have access to training data from a particular age range, this may restrict what our system requirements might need to be.

10.5.4   Communication Issues

The process of software engineering sometimes resembles the children's game "Telephone" – see Figure 10.4. In this game, children stand in a line and the first person whispers a message into the ear of the second person. They, in turn, whisper what they think they heard into the ear of the third person, and this continues until the message reaches the last person. Then the last person and the first person compare the original message and the final message (i.e. what the last person heard), which are often different in many interesting and amusing ways.

In product design, unfortunately, sometimes we have the same situation. The first person is the user. They communicate their needs to a member of the team (perhaps a

Figure 10.4 The "Telephone" game. Unfortunately, this is often an accurate description of the software process. Figure by Ellie Gabriel.

business analyst or a marketing person) who is tasked with identifying the needs of the users. This person in turn communicates these to the senior design team, who creates the system requirements. These are passed to the senior software engineers, who create a detailed design for the software and then pass it on to more junior software developers who are tasked with actually creating the software. The challenge is to ensure that what gets created at the end of this chain actually matches what the user said they needed and that we do not end up in the Telephone game scenario. The solution is to create detailed specification documents and to ensure that all key stakeholders review these such that the communication is not subject to "mishearing." There is no substitute for clear, written documentation in this process.

Clearly not everything that happens at each stage of the process gets communicated to the next stage. The needs-finding discussion, for example, will not be communicated to the system designers in its entirety. However, the relevant portions *must* be communicated, which puts a premium on the ability of the people in this chain to summarize the important parts of the preceding step such that the people down the line have all the information that is required to do their work. That is why it is important to provide background material in this document to explain "Why is it that we need to do this?" as opposed to just "What is it that we need to do?" One may want to think about the information that is passed down the chain as needing to satisfy the same requirement as a sufficient statistic. We can define this informally as "that summary of a data set that gives you all the information about the data set that you need for a specific task" – see Section 8.3. In this design scenario, think of the data set as the discussion that went on in the previous step and the sufficient statistic as the appropriate summary that the engineers in the next phase need to do their job.

Failure to communicate critical information down the chain can introduce significant problems and costs to our software product, and is a common cause of project failure.

There is dedicated software that many companies use to maintain the various documents involved and to keep track of the linkage from requirements to test results. One such software system is Helix [206].

10.6 Conclusions

In this chapter, we summarize the medical software life cycle process as detailed in the international standard IEC 62304 [119]. We note that while IEC 62304 appears to mandate a Waterfall process, this is not in fact the case and note that there are guides such as adapting Agile processes [1]. It is important to plan for the entire process first, before beginning the detailed work involved in any of the steps to identify potential issues that may arise down the road. You do not want to discover, for example, that you lack the ability to perform a validation study after you have invested significant resources in creating the software! As a final note: The formal medical software process often follows the creation of "informal prototype" software. Such prototypes can be used both to identify the needs of our users and also to allow an initial evaluation of the feasibility of the underlying algorithms.

10.7 Summary

- This chapter summarizes the medical software life cycle process. We will examine each component of this in more detail in Chapters 11–16. Please note that risk management is embedded at each stage of the process.
- The overall process is critically dependent on the risk classification of our software. Low-risk software receives lower levels of regulatory scrutiny. Moderate- and high-risk software have extra requirements for both the design phase (e.g. detailed software architecture design) and the testing phase.
- It is important to consider the entire process before beginning detailed work on any one aspect of the software. For example, if there are potential problems that might mean that the software cannot be validated (lack of resources) or deployed (e.g. dependency on App Store approvals), one should resolve these first, before beginning work on creating the software.
- An important aspect of the process is team management and communication. Information must be successfully transferred from project stage to project stage (and team to team) to ensure that the user's needs are reflected in the final product, and the final product is constrained to satisfy the user's needs (and nothing more!)

RECOMMENDED READING AND RESOURCES

This part of the book directly derives from the international standard IEC 62304. This is the definitive guide to the medical software life cycle process.

> *IEC 62304 Medical device software – software life cycle processes*, May 2006.

> *ANSI/AAMI/IEC 62304:2006/A1:2016 Medical device software – software life cycle processes (Amendment)*, 2016.

The following IMDRF regulatory guidance document describes the software life cycle process from a quality management perspective:

> IMDRF, SaMD Working Group. Software as a medical device (SaMD): application of quality management system, October 2, 2015.

One of the earliest descriptions of image-guided neurosurgery can be found in this paper, which was written just as the field was becoming established:

> W.E.L. Grimson, R. Kikinis, F. Jolesz, and P. M. Black. Image-guided surgery. *Sci Am*, 280(6):62–69, 1999.

NOTES

1. The use of Agile methods is no panacea, as illustrated in the case of the Healthcare.gov website described in the Vignette presented in Chapter 19 – that project used Agile methods in theory but not in practice.
2. This is the "Brown Cow" [216] model that we will discuss in more detail in that chapter.
3. It is also worth remembering that not all software engineers are native speakers (of, for example, English in the United States). Keeping the language simple and consistent will help with this language barrier also.
4. One additional advantage of writing a validation plan in this way is that it conforms to the structure commonly used in NIH grants, and as such one could reuse most of this as an Aim in such a proposal. This is particularly relevant to small businesses (in the United States) who qualify for Small Business Innovation Grants (SBIRs).
5. This manifests itself primarily in how detailed some of the design and verification and validation documentation/designs/plans need to be.
6. This three-class classification is also found in the FDA regulations for medical devices – see Section 2.3. Conversely, the latest IMDRF guidance (see Table 2.1) has a four-class categorization. What is common in all of these cases is that the level of scrutiny, and by extension the level of detail in the documentation and the design, increases as we move from software that cannot cause serious harm to that in which misuse/errors can cause serious injury or even death.
7. See Chapter 5 for definitions of some of the terms, such as risk and hazard.
8. There are some interesting changes between the original edition of the standard [119] and the amendment [2] that reflect the changes in the consensus as to how software should be developed. Some of it is purely terminology, such as the change from "software product" to "Medical Device Software." Some of it is more substantial

such as comments that read "Replace '[Class B, C]' with '[Class A, B, C],'" which effectively expands requirements that were previously only for Classes B and C to Class A as well. The take-home lesson is that the standards evolve over time and one needs to stay up-to-date with changes as they happen.

9. The GPSV separates system and software requirements, where the system may be considered as a device that has both hardware and software. For this introductory analysis we assume that our software is the system, that is, we operate in an SaMD environment.

10. Consider the case of software that is designed to run on an iPhone. If we cannot install it on devices as it would violate Apple's guidelines and hence cannot get approved by Apple, we may need to reconsider our project.

11   Identifying User Needs

INTRODUCTION

This chapter discusses the processes of identifying the needs of our users, which is the first step in the software life cycle. We begin with an overview of the structure of creating new projects. Next, we apply the "Brown Cow" model (Section 11.1) as a guide for how to go about understanding the user's current situation and future needs. We will then touch on the formal process for creating a use specification as discussed in the guidance document IEC 62366-2 [124] (Section 11.2). Finally, we "begin" our example image-guided neuro-navigation project (Section 11.3) and use this to illustrate the concepts presented in this chapter. We will return to this project in the next few chapters as we continue along our software life cycle journey.

Yock et al.[1] divide the process of creating new medical technologies into three interlocking phases:

1. needs finding and needs screening;
2. concept generation and concept screening; and
3. strategy development and business planning.

In the context of this textbook, we will make the important assumption that the "big picture need" (e.g. curing cancer) has already been identified. Hence, we assume that somebody has already identified a significant clinical problem that can be solved using new software. This may involve both a series of unstructured, open-ended, user observation sessions and more formal questionnaires and structured interviews – see Section 11.2.

At first brush, therefore, our problem is solved, our needs have been identified, and we are ready to start implementing our software. However, there is a huge gap between a "conceptual big picture need" and the actual detailed needs of a user of a piece of software. In our context, "need" does not mean "curing cancer" but rather "the functionality needed in our software so that our clients can use it to cure cancer." Bridging the language gap between a clinical need and the software needed to create the appropriate functionality is an art in and of itself, and this section attempts to provide an overview of how to accomplish this.

A Note on Regulations Any software that will be used in a clinical setting may need to satisfy regulatory constraints. As a reminder, the FDA defines its mission as assuring the safety, efficacy, and security of medical devices (in this context). Always keep these three words in mind. As you proceed with your discussions, ask how one would demonstrate efficacy (the software can be shown to do what it claims to do), safety (it does not expose the patient or user to undue risk from its use), and security (the software was designed to protect patients and users from unintentional and malicious misuse). Data privacy issues should also be addressed, if applicable.

11.1 Exploring the User's Needs

11.1.1 The "Brown Cow" Model

Suzanne and James Robertson [216] present a model of how to begin this process of needs identification. They label this the "Brown Cow" model (named from a children's rhyme), which divides the process of identifying the problem into four parts:

1. *How Now?* – model of the current situation;
2. *What Now?* – essential business use case;
3. *What Future?* – enhanced business use case;
4. *How Future?* – product use case scenarios.

This is a taxonomy of the problem that can serve as a useful roadmap for the overall process. The "How Now" step aims to understand how the problem is currently addressed in a tangible sense. From here we can abstract the current situation to its essence, which is the "What Now" stage. To use a simple example, "How Now" may be clearing snow after a storm, while "What Now" is simply the task of keeping one's driveway accessible.[2] "What Future" then asks what else we want to be able to accomplish at a high level (perhaps also add constraints to how much time it takes to do the procedure), which then leads to "How Future," the process of improving the current equipment or perhaps adding under-driveway heating elements to melt the snow.

11.1.2 Understanding the Problem Area

Once the main area of discussion is identified, and before an initial meeting with clinical collaborators, clients, mentors, and other partners, the software engineer needs to be educated about the problem at hand. We suggest that the first step is to ask for (if possible, otherwise find) reading materials that describe the underlying clinical problem and the current approaches to solving it. This is important as we need to acquire the vocabulary and understand the "geography" of the problem. For example, if the problem area is brain cancer, we need to learn something about how the brain works and something about brain cancer as a disease. Next, we need to understand something about the current procedures for diagnosing, managing, and treating the disease, who the people (oncologists, radiologists, etc.) involved with this process are, and what the process is. Finally, we need to have a sense of what the patients typically are like (age, sex, obesity, other diseases, etc.) as this may also be a factor.

The key to a successful first conversation is the establishment of a common language between the participants. Hence, you as the technical person need to acquire some core vocabulary about the disease to enable you to understand the context in which the problem is set. This includes some underlying anatomy and physiology. Remember that while ideally both you and your clinical partner will need to "cross over" and understand what the other does to allow for successful completion of any project, in practice you will not meet in the middle, but closer to the clinical side.

An important part of the preparation for the first meeting is to make a list of issues that appear confusing about the problem and have these clarified. It is often useful to send a list of questions (keeping these on the short side) ahead of time to help your mentors orient themselves as to what you know and what you need to learn.

As we will re-emphasize later, it is imperative to not make any assumptions about the possible solutions or how to solve the problem at this early stage. Instead, the engineer should come up with a series of cogent questions to prime the discussion.

A direct meeting with a client is typically a great way to understand the "unmet needs" of the system and to understand the "pain points" of the current system. However, there are additional critical items that the developer has to keep in mind that would not be covered in this meeting. For example, typically a client will rarely articulate things that are basic needs ("known knowns") but are important for a good product nevertheless (e.g. the client may expect that if the software crashes any work done up to that point should be retrievable and the physician won't have to restart the procedure from scratch). While the customer may never have communicated this need, the absence of a basic given such as this would likely be a point of bother later and perhaps even a competitive disadvantage. Another category that the developer should be wary of is the "unknown knowns" from the perspective of the customer – for example, the software pops up, perhaps quite logically in the eyes of any developer, a warning dialog every time a user takes a certain action, but the constant pop-up becomes an annoyance to the physician. Perhaps there is a better way/better design to present the warning to the user that the developer should think about. A client may also not communicate features that may make the software innovative and add competitive value to the product (e.g. being able to transfer to and view surgery images from your smartphone). Understanding such types of needs and the exciters/innovation points of the system may also be helpful for the software designer.

11.1.3   The Initial Meeting

Almost all projects start with an initial meeting that may include a number of stakeholders, such as the clients, managers, software designers, and business analysis. This is your opportunity to meet your future collaborators and establish a good working relationship. The usual rules of making a good first impression apply here. You need to be on time and dressed appropriately for the environment you will be in. If in doubt, ask. Also, there may be many people in the room besides your primary mentor/contact. Try to figure out who is in the room, their background, and their interest in the project. In addition, it is helpful to create an agenda and make sure that everyone gets a copy of this before the meeting.

One thing to remember during the first meeting is to resist arriving at any conclusion too fast. To somebody with a hammer, everything looks like a nail. To a database programmer every problem looks like a database design problem (at least at first). Keep your eyes and ears open and remember that there are two types of things you do not know, the so-called "known unknowns" and "unknown unknowns." Known unknowns are the things you know you do not know. In some respects, these are easy to address you can simply ask questions and research the topics. Unknown unknowns are harder. If you are not even aware of a potential issue, then it is hard to plan to address it. A way to deal with this in our situation is to ask questions like "Is there anything else I need to know?" or, perhaps, "Would you like to add anything?" The key, as always, is to *listen* and make sure you understand. It is better to ask more questions than to pretend that you understand what is going on and pay the price later.

Another method to ensure you understand the problem statement is to regurgitate the summary statement in your own words to the entire group. This ensures that all the concerned parties are on the same page.

Try to understand what the problem being presented really is. Sometimes what the clinician feels their problem is, is not the actual problem but a consequence of some secondary issue that is easy to resolve. This is moving from "How Now" to "What Now." To follow the previous metaphor, try to move your collaborators from stating "I use a shovel" to "I need to keep my driveway clear." This is the essence of the problem. From there you can begin to understand what the real issues are and how to address them. Here is a (partial) checklist:

- Try to resist reaching a solution too early. You may not understand the problem.
- Do not let the clinician's idea of the solution necessarily bias what the best solution is. They also "don't know what they don't know."[3]
- Do not assume that the current way of doing things is necessarily bad as a result of their "technical incompetence." You may think you have a better way, but there may be very good reasons for some of the current practices. Ask and try to identify what is critical and what is accidental. Some of the current methods may also be constrained by legal/regulatory issues.
- Identify the lay of the land. What comes before this procedure and what comes after? What happens to the patient prior to this point and what will happen after this treatment? Where is the "data" coming from? Where does my output need to go? What constraints does this impose on the problem?
- Think about the time limitations of the process. Is this a time-limited process? Are there things that we can do ahead of time?
- Ask about safety/privacy issues and any other limitations in terms of what equipment/computers can be used and what cannot.
- Ask also about regulatory/safety issues that may govern aspects of the problem.

- Ask to observe a procedure or two if possible. This is vital in enhancing your understanding of the overall situation ("How Now").

Remember also that many physicians and providers can be 'headstrong' characters with very strong opinions. This, perhaps, is an occupational hazard as it is part of the nature of exuding confidence to the patient and their families. However, such individuals are not necessarily correct in the assumptions they make or their understanding of the proposed solutions that may work best. "I just want something that does this" may simply mean "I want something better than my colleague across the street." Our advice is that one should try to figure out what the users really need as opposed to what they think they need. In fact, it is often stated in human-centered design that a person may say they "need" something but the better plan is to ask them to show you why or how they would use it. This is often much more illustrative and often results in a better understanding of their true needs.

It is always a good idea to ask explicitly about how one would convince them that a proposed solution works. Phrased differently, perhaps, what evidence would it take for them or one of their colleagues to buy your software?

There may also be other stakeholders (especially in the deployment process) who have important needs that become important influencing factors on what suite of software products get installed in the clinical environment. The needs of such stakeholders may dictate aspects of the development and need to be identified from the outset. For example, clinic managers may require integration with specific EHR products (Section 3.3.1), IT may require specific cybersecurity features (Section 3.5), and distributors may require ease of installation and cloud-based upgrades. One needs to be aware of these stakeholders, who may not appear to be "users," but also have "needs" that need to identified and satisfied.

In summary, prepare thoroughly, read ahead, listen, and ask lots of questions. Do not worry about coming up with a solution, but simply keep an open mind. Ideally, at the end identify some goals for you to work on and schedule the next meeting. Clinicians are busy people taking care of patients daily, and it will behoove the software engineer to have a set of questions ready for the meeting. Clinician schedules are also somewhat unpredictable and any meeting may need to be postponed at the last minute as a result of surgery taking longer or some other emergency, so one needs to be flexible. Sometimes scheduling meetings very early in the morning is the best strategy.[4]

At the end of the meeting, briefly summarize action items and, in particular, who is responsible for what and the expected timeframe to complete the actions. Also, soon after the first meeting, try to summarize what you have learned (in writing) and identify "holes" in your understanding. Then try to follow up with more questions.

Note: These meetings are not part of the regulated design process. They are in some sense (see Vogel [261] again) "research" as opposed to "development." The outputs of these discussions form the inputs to the regulated design process that you will need to follow to design, verify, and validate your software. This is what the Quality System Regulation (QSR) [69] refers to as design inputs. These discussions will also form the basis for the creation of the "use specification" document, discussed in Section 11.2.

11.1.4 Some Additional Checklists

In any biomedical project the following issues are worth considering:

1. Make sure you understand why the problem presented is significant. Why should anybody (or really lots of people) care about a solution to the problem?
2. What has been tried before to address this and why are these alternatives not satisfactory? Is the primary driver quality or affordability?
3. What are the external constraints to the solution (external environment, interfacing issues, time, operator skill, anything else)?
4. What resources/skills/time commitment will it take to complete this project?

Robertson and Robertson [216] begin their book with a short section discussing 11 items (labeled "truths") about many software projects, and these are an excellent complementary checklist to the one above. Please note that while most of these apply to medical software design, one in particular (number 5) does not (at least if interpreted literally). We present these here as pointers for discussions and as an additional checklist:

1. Requirements are not really about requirements.
2. If we must build software, then it must be optimally valuable for the owner.
3. If your software does not have to satisfy a need then you can build anything. However, if it is meant to satisfy a need, then you have to know what that need is to build the right software.
4. There is an important difference between building a piece of software and solving a business problem. The former does not necessarily accomplish the later.
5. The requirements do not have to be written, but they have to become known to the builders.
6. The customer won't always give you the right answer. Sometimes it is impossible for the customer to know what is right, and sometimes he just doesn't know what he needs.
7. Requirements do not come about by chance. There needs to be some kind of orderly process for developing them.
8. You can be as iterative as you want, but you still need to understand what the business needs.
9. There is no silver bullet. All our methods and tools will not compensate for poor thought and poor workmanship.
10. Requirements, if they are to be implemented successfully, must be measurable and testable.
11. You, the business analyst, will change the way the user thinks about his problem, either now or later.

Item 10 is particularly important for medical software design. We will return to this again in the next chapter (on writing requirements). However, any time you hear words like "should work well," "should be easy to use," or "should be accurate" describing user needs, you should always try to create/identify an explicit quantitative metric for evaluating these to serve as a testing criterion.

11.2 Creating a Use Specification Document

The international standard IEC 62366, titled *IEC 62366: Medical devices – application of usability engineering to medical devices (Parts 1 & 2)* [122, 124], provides important guidance as to how to apply techniques from usability engineering or human factors engineering to the design of medical devices (and software) – this is defined as "application of knowledge about human behaviour, abilities, limitations, and other characteristics to the design of medical devices (including software), systems and tasks to achieve adequate usability" [122]. Naturally this leads to a definition of usability: "characteristic of the user interface that facilitates use and thereby establishes effectiveness, efficiency and user satisfaction in the intended use environment."

Usability engineering provides important techniques both for identifying the user's needs and for optimizing the design of our software to better meet them safely (i.e. risk management) and well (hopefully leading to commercial success). An important part of the process of capturing the needs of the users is to create a "use specification document" [124]. As the guidance states, this is the foundation for defining the use. We present some initial pointers as to how one would go about creating such a document.

11.2.1 Users, Uses, and Environments

The first step is to analyze the "intended users, anticipated user tasks and intended use" [124]. These are both critical for preparing the system specifications document – see Chapter 12 – but also because a good understanding of these can help identify who we need to talk to and what places we may need to visit, so as to better understand the needs and constraints of our users. This will also help us to later create the appropriate user interface.

Users These may range from patients or lay caregivers, to technologists, physicians, and many other types of people. For each type (or group) of user we create a user profile that includes characteristics such as:

1. demographic information (e.g. age, education);
2. knowledge and skills (e.g. language proficiency, computer skills);
3. limitations (e.g. visual impairments); and
4. tasks that they need to perform.

These profiles relate to what is often called a "persona" [37]. This is a detailed description of a fictional user that often helps to make things concrete. One can always ask, for example, "What would Dr. Jones do in this situation?"

One must be careful not to over-generalize users. For example, we may think that one of our users is a radiologist. However, a radiologist at a major academic center may have very different training, experience, and needs compared to a radiologist at a community hospital. In this case, therefore, we should treat these two radiologists as different users with their own, no doubt partially overlapping, needs.

User Tasks This can be developed by examining "comparable and predecessor" software. What we need here is a sense of how the users will use our software.

Use Environment Where will our software be used? What constraints does this create? We discuss this more in Section 12.3.3. It is important to try to visit such environments so as to have a direct understanding of the situation that a user is in. This will also help with the "known knowns" (i.e. items that a user will never articulate but which are critical). For example, surgeons wear gloves during operations. This is obvious knowledge to them but perhaps unknown to a software designer who has not been involved with surgical procedures before. The wearing of gloves may impact how the user interface needs to be designed.

11.2.2 Techniques for Developing a Use Specification

IEC 62304-2 lists a number of different methods that can be used. We list them below:

- contextual inquiry and observation – an interview at the workplace;
- interview and survey techniques – additional interviews with users either individually or in groups;
- expert reviews – asking domain experts to review our designs/task lists;
- advisory panel reviews – this can be a group of 6–12 people who review the process;
- usability tests – once we have prototypes.

These techniques can be used both at the beginning of the process (e.g. establishing user needs) and during the process to evaluate proposed software designs (especially the user interface). We will continue this discussion in Section 13.5.

11.3 Example: The Image-Guided Neuro-navigation Project

For our fictional neuro-navigation project – see Figure 10.1 – we first must allow ourselves to imagine that it is the year 1995 (i.e. about the time this type of technology was being developed). We first meet with our key clients (our neurosurgeons) and listen to them to understand what it is that they do. We may want to get some information to answer the following questions (and any others that may appear): What is the disease (brain cancer in this case)? How do patients get to the point that they need surgery? What happens prior to the surgery? What happens after the surgery? What are the pain points in this process? What is the current cost of the procedure? What do current solutions (if they exist) cost? What would somebody be willing to pay for something better? How many people will need something like this? What can be done to ease the burden on the doctors performing the procedure? Whenever possible, send clients questions ahead of time.

The next step is to visit (ideally many times) and observe the process in action. Let's go back in time and imagine it is 1995. What is the situation in the neurosurgery operating room? What we probably see in the room is medical images such as MRIs or CT images displayed on lightboxes (see Figure 3.2) and even printouts of surgical plans attached to the wall with tape. As she operates, our surgeon moves back and forth between working on the patient and looking at the images on the wall to orient herself as to where she is at any given time. This requires significant experience in these procedures as all mapping of the images to the patient is done mentally.

What is our neurosurgeon trying to do? She is trying to remove as much of the tumor as possible, while (1) sparing as much normal tissue as possible; (2) avoiding functionally important zones (e.g. the motor areas of the brain, which if damaged will lead to partial or full paralysis); and (3) complete this procedure in the minimum amount of time possible.

What would our neurosurgeon like to do in the future? She would like to be able to integrate knowledge of functional imaging (e.g. functional MRI) to improve her ability to avoid functional zones. She would also like to be able to do her work better and faster, as right now moving back and forth between the operating table and the wall to look at the images is tedious, time-consuming, and challenging, as she needs to keep all this information in her head at all times.

How will a new product help her? Ideally, we will have a piece of software that shows her the location of her surgical tool with respect to the images of the patient acquired prior to the surgery, as well as the trajectory of the incision and dissection to get to the tumor. This will eliminate her need to keep going back to the wall to look at the images and will reduce the mental stress of keeping all this information in her head at all times. The software should be able to import functional images to define zones to avoid, and it should also include the ability to create surgical plans to guide the procedure. This will have real-world implications as she might be able to perform a better surgery with more success and complete removal of a brain tumor. This might be life-saving and it might also have more complications down the road.

This is hopefully the type of information one should extract from these initial conversations so as to be able to begin designing our system.

11.4 Conclusions

The process of identifying user needs is one of the most challenging aspects of creating medical software. First, one must identify who the users (and other stakeholders) actually are. These may include the obvious (e.g. doctors), but also IT specialists who need to interact with and integrate the software into a larger workflow. Next, we need to understand what the users need to do. This involves an understanding of the clinical situation (e.g. potentially the disease that is being treated) and the environment in which they operate (e.g. the operating room) and the constraints that this imposes. This is a critical and often time-consuming process that may require multiple meetings and repeated visits to clinical facilities.

11.5   Summary

- We presented the Brown Cow model for understanding the needs of your users. How are they solving their problem now? What are they trying to do? What else would they like to be able to do? How will they be able to do it? There is a creative tension between the concrete ("How") and the abstract ("What") that permeates the whole process.
- Understanding the needs of your users may require you, the software engineer, to invest significant time in learning about the underlying disease, the clinical environment, and a myriad of other issues that are involved.
- It is important to try to identify what the "unknown unknowns" (from the perspective of the software designer) are. Sometimes these are known to your clinical collaborators, but they are so obvious to them that they neglect to mention them.
- While identifying the needs of your users, resist the temptation to arrive at solutions. Focus instead on the conceptual big picture.
- There is no substitute for observing the user in action, in their actual working environment.

RECOMMENDED READING AND RESOURCES

Identifying user needs is a complex process. A good introduction to the topic can be found in the book:

S. Robertson and J. Robertson. *Mastering the Requirements Process: Getting Requirements Right*, 3rd ed. Addison-Wesley Professional, 2012.

The core international standard on usability engineering is IEC 62366, which comes in two parts (part 1 is the standard and part 2 is a guidance document):

IEC 62366-1 Medical devices part 1: application of usability engineering to medical devices, February 2015.

IEC 62366-2 Medical devices part 2: guidance on the application of usability engineering to medical devices, April 2016.

SAMPLE ASSIGNMENT

We assume here that students are divided into small groups (3–4) and are each assigned to work on a project that is proposed/supervised by an external (clinical) mentor.

1. Please introduce yourselves to your mentor and make arrangements for a first meeting. I suggest each student write a brief bio (one paragraph) and that a single student collate these to forward to the mentor. It is best to include the instructor in this email chain as he or she should be present at the initial meeting.

2. Ask your mentor for background reading materials as to both the clinical problem (e.g. the disease, current treatments) and the open challenges in this area.

3. During the initial meeting, follow the suggestions from Section 11.1.3 to try to identify what the situation is. Listen carefully and do not pretend that you understand things that you do not.

4. Try to schedule a follow-up meeting that ideally lets you visit the environment in which the project will take place and observe the process. For example, if this is a project for software that will be used in an operating room, visit the operating room and observe the process "as is." This will help you understand how the process is currently done and what the challenges are.

5. Create a shared document to collect your notes and thoughts through these meetings.[5] Take pictures of places (if allowed; hospital environments often do not permit this) and drawings (if a whiteboard is used) to help document these.

In addition: (1) Please remember that your mentors are busy people. Do not inundate them with emails. Instead, have one member of your team (this can rotate each week) be the point person responsible for communicating with your mentor. Try to offer times to meet that work for everybody. Remember that busy people do not read long emails, so try to be concise and professional in all of your communications.[6] (2) If you need to visit a hospital, try to dress somewhat professionally (patients will see you as staff). Ensure you have ID cards displayed if this is required. In recent years, entry into the hospital and medical settings is becoming much more restricted to protect patients and their privacy, so you may need to coordinate your visit. Try to arrive early as hospitals are often mazes, so you have to allow for time to be lost and to find your way. In addition, hospitals often use credentialing systems (e.g. Vendormate, Sec³ure). You may need to reach out to the clinical manager (or whoever else is managing your visit) well ahead of the intended procedure date to determine if they require such clearance.

NOTES

1. The website of this book has some very interesting videos and other materials.

2. There was a blizzard going on outside at the time these lines were being written.

3. At the time of writing (2020), many discussions come to a point where a clinician will say "I need an AI to do this." While AI/ML methods may be appropriate for some problems, there are many, many problems, where they are either completely inapplicable, or for which traditional techniques can work just as well, if not better.

4. As mentioned before, it is important to create and distribute an agenda ahead of time. Include time for introductions and for everyone to explain their role. Make sure everyone understands the project goal and the meeting objectives. Include discussion points of the key topics you need to cover.

5. Our students have discovered that Google Docs is an excellent platform for this task as it allows for easy sharing and simultaneous editing.

6. Imagine a busy person looking at your email on their phone. They will, most often, simply glance at the message and never scroll past the first page. One should try to include all critical information in those first few lines that appear on the screen, so that when the email is first opened it will be seen! For example, if you are requesting a 4 p.m. meeting for tomorrow, state this first and then explain why. If the person receiving this is booked for that time, you may get an immediate negative response which allows the conversation to proceed. If, instead, the email begins with a long explanation, your correspondent may not have a chance to process the information until much later in the day.

12 The System Requirements Specification

INTRODUCTION

This chapter describes the process of creating a system requirements document which is one of the most important steps in the software life cycle. First, we orient ourselves as to our position in this life cycle (Section 12.1). Next, we present a brief review of key regulatory issues (Section 12.2), and following this we describe a template for creating this document (Section 12.3). Next, we discuss in more detail the process of writing a requirement (Section 12.4). As always, we then revisit risk management (Section 12.5) and address how to evaluate our requirements for potential harms. Section 12.6 discusses how to review system requirements. The chapter concludes by continuing our discussion of the example image-guided neuro-navigation project by presenting an outline of what the system requirements document for this might look like (Section 12.7).

12.1 A Bridge from User Needs to Software Design

We are now at the first step of the formal life cycle process the identification of system requirements. The approach presented in this chapter is not, by any means, the only way to do this, but *one possible method.*

The system requirements document forms a bridge between the informal "discovery" phase of a project in which we identify user needs and the underlying issues in the work to be contacted and the actual design of the software itself. It is the input to the next document in the chain, the software design document, as well as to the validation specifications (or validation plan). The goal of the system requirements is to capture all necessary information from the discovery phase in a structured format so that the creators of these subsequent documents have all the information they need to perform their work. The lack of a proper requirements process [33] can lead to disaster. See, for example, the case of the Healthcare.gov website described in the vignette presented in Chapter 19.

This document is also addressed to the actual users as a way to demonstrate that their needs are addressed by the design. They will also need to review this to ensure that there are no omissions. This, along with other design documents, aims to formalize our understanding of what the user needs. While an informal understanding may have been arrived at during the initial brainstorming sessions, these oral agreements can be ambiguous and often hide misunderstandings, such as that different discussion participants can mean slightly (or sometimes very) different things by the same words. A formal document can help clarify to the end user what the design team's understanding of the problem is.

Unlike in a class project, in a large organization each of these documents is created by separate (though often somewhat overlapping) teams, so each document must contain enough information to enable the recipient to perform their work as needed.

Note: While the idealized Waterfall model assumes a completely sequential process, this is rarely the case in practice. Once the system requirements document is completed, the software designers will use it as an input to create software specifications (and similarly the testing engineers to create validation plans). At this point, they may (will) identify ambiguities and omissions in the system requirements that prevent them from performing their task. The result of this is that the software design stops and we go "back up" to the previous step to revise the system requirements to address these concerns prior to resuming the design of the software. The key to a controlled process (mandated by the FDA quality regulations [69]) is not that it has to be sequential, but rather that we know in what stage in the process we are at any given point (e.g. system requirements vs. software design).

12.2 Regulatory Background

A good place to begin here is the formal definition of the terms *requirement* and *specification*. These are defined by the FDA as [70] (Section 3.1.1):

A requirement can be any need or expectation for a system or for its software. Requirements reflect the stated or implied needs of the customer, and may be market-based, contractual, or statutory, as well as an organization's internal requirements.

... A specification is defined as "a document that states requirements." ... It may refer to or include drawings, patterns, or other relevant documents and usually indicates the means and the criteria whereby conformity with the requirement can be checked.

The whole requirements process is well summarized in the IMDRF QMS document [127] as:

This is a customer-driven process that requires clear, and often repeated, customer interaction to understand the user needs. These user needs are then translated into requirements. Well-documented requirements can then inform the testing activities later in the design cycle. There are other sources of requirements that can include regulatory or non-customer specified performance requirements.

As the quotes above make clear, the requirements do not come purely from user needs. For example, a software designer also needs to be mindful of applicable data security and privacy regulations – see Section 3.4 – and ensure that their design is compliant. One should consider these as additional safety and security constraints, which will result in additional non-functional requirements. The same applies to cybersecurity considerations – see Section 3.5. There may also be regulations in the specific area of interest and also conventions that must be followed. Again these conventions may result in extra nonfunctional requirements.

Finally, the IMDRF QMS documents then reminds us that the requirements may change as the process evolves and the developer gains a better appreciation of both the user's needs and their environment.

12.3 An Example System Requirements Template

The system requirement document is a structured document consisting of various sections. We present here a simplified adaptation for the setting of an undergraduate class project. It is intended, however, to be relevant to the needs of real-world products.

Our template has the following outline:

1. Cover and revision control pages;
2. Introduction
   a. goals and objectives
   b. statement of scope
   c. definition of terms
   d. references
   e. significance;
3. Overall description
   a. intended users
   b. intended use
   c. intended use environment: physical, standards, legacy environment, platform;
4. Functional requirements;
5. Non-functional requirements.

12.3.1 Cover and Revision Pages

The system requirement document is a key component of the formal design process. It is a versioned document. All approved (and *signed*) versions of this document must be saved in what is called the "design history file" [69] of the project, which tracks the history of the project requirements. This may literally involve scanning these signed documents to create a digital version.

The cover page of the document (Figure 12.1) contains the usual cover page information, including the project title, the authors' names, and the current version and date. The second page (or more pages) contains the version history of the document. Each time a new version is approved (at a design review meeting), it is added to this table (which may run to multiple pages). The version number is arbitrary (much like in software) but should follow an ascending order. The comments field should summarize the major changes made from the previous version (in as much detail as would be useful). There will also often be an "approvals" section showing the signatures of the authorizing managers.

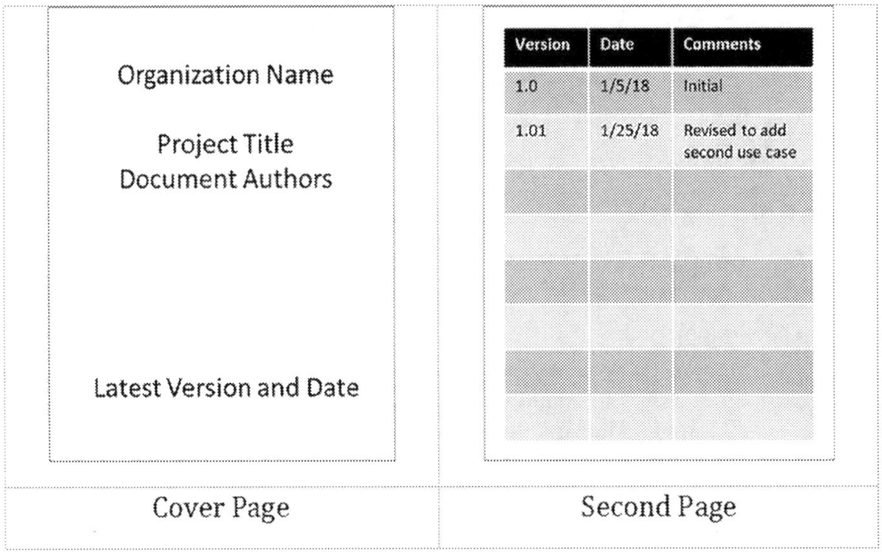

Version	Date	Comments
1.0	1/5/18	Initial
1.01	1/25/18	Revised to add second use case

Organization Name

Project Title
Document Authors

Latest Version and Date

Cover Page   Second Page

Figure 12.1 Cover and log page for a versioned document. In a versioned document, such as the system specifications, it is important to have both an explicit title page and a log page listing the revisions and approval dates. Missing in this simplified example is an approvals signature section, where the authorized signatories sign to confirm approval of the document.

12.3.2 Introduction Section

This section introduces the project and the document. The first two parts (goals and objectives and statement of scope) should be brief and to the point.

The significance component of this section is sometimes not found in this document. We believe that this is an important section for all projects, however, as it provides some big-picture information for the software and validation teams (who need to write the software and validation requirements documents) to enable them to better appreciate why the system requirements are what they are.

a. Goals and Objectives: This is a big-picture description of the project. We define here what the problem is that we are trying to solve and how (in very high-level terms) we are going to go about solving it.

b. Statement of Scope: This should consist of a brief description of the software, including the inputs, the outputs, and any major functionality that will be included. One should avoid discussing any implementation details. As part of inputs and outputs, one should also consider interconnectivity. One must outline what outside/hospital systems this has to connect to (e.g. PACS, EHR systems) and what standards will need to be implemented for communications (e.g. DICOM, HL-7 – see Section 3.3).

c. Definition of Terms: This is a glossary that contains the definitions of key terms.

d. References: This will include a list of the key applicable regulatory documents and any other important sources. This later category might list, for example, papers describing a technique to be implemented or a pitfall to be avoided.

e. Significance: This section addresses the "Why should I care?" question. This is the first and most important question in any project. Another way to think about significance is to ask the question: "If this project is successful, what is the potential impact to both society at large, and from a commercial perspective?" To summarize this further, we get to: "Will this make a difference?" If the answer is negative, the project is a waste of time. The significance section can be structured as follows:

First, we present a review of the underlying disease with some statistics in terms of numbers of patients, cost of treatment, and other relevant issues. This is necessary as some of the readers (e.g. computer programmers) need to be educated as to the medical background of the actual problem.

This is followed by a review of the specific clinical problem that our product will address. For example, in the case of planning a plastic surgery procedure, we move here from details as to the causes of the problem that require plastic surgery to the actual current practice of the procedure itself. How is this planned? What tools are currently available for this?

Next comes a critical review of the current tools focusing on key weaknesses and their consequences. Lack of ability A causes problem B (e.g. longer procedure times or the need for more physician training). It might be that the technology exists but it is not available in a cost-effective or user-friendly manner.

Finally comes a statement as to how our proposed solution will help and how it will impact patients and be a success commercially.

12.3.3   Overall Description

Intended User(s): Who is the intended user? To follow up the previous example, "The intended user is a trained neurosurgeon or neurologist in a large academic hospital." Defining the users explicitly allows us to, among other things, better decide on the level of sparseness of the user interface and the documentation. This is the *Who* from the Clue game (Figure 10.3).

Intended Use(s): What is the high-level task that the software will perform? An example of a use case is: "This software will be used to measure the volume of gray matter from high-resolution brain T1-weighted MRI images." Please note that this definition answers the "what" question but not the "how" question. How the software will do this may not necessarily concern us at this point: the key here is what the "customer" of the software would like to do at a high level. List these intended uses, one after another (if there is more than one). This is the implicit *What* from the Clue game analogy. (In Clue, of course, the what is always "murder.")

Intended Use Environment(s): This is the "where and with what" (both in terms of location and environment) question. Within this we can distinguish several sub-aspects of this topic [261]:

- Physical environment: Will this be used in an operating room, office, ambulance, in the field, or anywhere else? Will the users be gloved? Is the room dark or bright? Anything else?

- Platform: Will this run on MS Windows, macOS, iOS?
- IT environment: Here we answer questions such as: How does our software fit with the rest of the user's workflow? What format does the data come in? How? Where is output going to be stored? Is there a database server that we need to connect to?

In addition, we have implicit constraints arising from regulatory and legacy issues such as:

- Standards: We need to be aware of whether there are task-specific regulatory standards that our software needs to comply with.
- Legacy: Is this a replacement software for something that already exists? Do the users have certain expectations that we cannot violate? For example, radiology workstations display images such that the left of the patient is on the right of the screen. This is what the users expect as this is what current (legacy) systems use. Changing this will lead to confusion and medical errors.

In general, a medical software package does not solve a problem end-to-end, but rather becomes part of a clinical workflow. It is *critical to identify how our solution will fit in these proceeding/succeeding steps* (e.g. Are inputs coming in as files? If yes, we also need to identify the file formats).

It is a good idea to number your uses, users, and environments so that you can reference them later.

The intended use(s) (and also intended users and environments) limit the marketability of the product. The statement of intended uses provides the functionality of the device. If this statement is modified it would require the company to resubmit to the regulatory body for clearance. This is also the same for intented user and intended environments

12.3.4 Functional Requirements (Efficacy)

At this point one should list one-by-one the requirements for the system. These are the things that the system must be able to do to make it useful. Useful here means to satisfy the overall goals of the project, which ultimately satisfies the needs of the users within the constraints of the regulatory environment.

12.3.5 Non-Functional Requirements (Safety and Security)

These are requirements that are often implicit in the users' minds, but assumed to be there, and/or are required for regulatory clearance whether the user is aware of this or not. Into this category fall: (1) safety requirements how to ensure that the system will not cause harm to the patient or the user; (2) security requirements how to ensure that the system cannot be used (maliciously) to cause harm. For software this may mean preventing unidentified access and/or malicious alteration of patient data by unauthorized users, and ensuring patient confidentiality, among other things. In addition we may need to consider (3) data privacy and cybersecurity requirements to ensure protection of sensitive user data.

A separate set of non-functional requirements answer the "how well" question. As an example, let us revisit the use case of computing gray matter volume. The functional requirement(s) here may well be specified as accuracy level and possibly computational time. The non-functional requirements may have to do (in addition to safety and security) with ease of use and the intuitiveness of the user interface. This is both a risk management issue, as a good user interface may prevent user errors – see Section 13.4.3 – and also good business strategy. For example, non-functional requirements may arise from the business needs of the organization, such as creating something that has emotional appeal for marketing purposes – see also the discussion in Section 6.2.[1]

Non-functional requirements are critical. In the case of the Healthcare.gov website, which is described in the vignette presented in Chapter 19, one of the sources of difficulty was the lack of consideration of non-functional requirements such as quality and security.

12.4 Writing a Requirement

The requirements must be written in such a way that the engineers tasked with designing both the software and the validation plan can both understand them and act on them.

Good requirements need to be "complete, accurate, unambiguous, traceable and testable" [261]. The first three are mostly obvious. Traceable means that the requirement can be traced directly to specific use cases.[2] Testable means that the requirement as written can be used to generate a test that can objectively demonstrate that the system meets the requirement. For example, a requirement that "the system should be able to perform the task quickly" is not testable as "quickly" is not specific enough. It becomes testable, if, for example, the requirement is rewritten as "the system should be able to perform the task in less than 30 seconds." Then, at the validation phase, we can confirm by "examination and provision of objective evidence" that the particular requirements for a specific intended use can be consistently fulfilled, that is, we can measure how long it takes a user to do a task and declare success if it takes them less than 30 seconds.

Finally, a requirement at this level must be somewhat abstract and refrain from overly constraining the software design unless this is absolutely necessary. For example, one would not specify type font size for the user interface in a system requirement unless, for example, this is deemed absolutely necessary because of the class of users involved. Otherwise, the system designer should restrict himself or herself to specifying what the necessary visibility of the screen should be (e.g. easily readable from a distance of 5 feet) and let the software engineer determine what the appropriate solution is.

Stylistically a requirement is written using the "shall" form [261]. This marks it as being non-optional after all, this is something that the system must do to satisfy the needs of the user. For example, we can write something like: "The software shall compute the volume of gray matter in the brain from high resolution T1-weighted MRI images with an accuracy of greater than 95 percent and in less than 25 seconds."[3]

The use of "shall" makes it clear that this is a requirement and the quantitative markers at the end (95 percent accuracy, 25 seconds) enable the validation team to create a set of tests to demonstrate that software does what it is supposed to do.

The following is a template for writing a requirement:

1. The actual requirement: "The system shall . . .".
2. User/uses/use environment in response to, e.g. this requirement relates to User #1, Uses #2 and #3, in environment #5.
3. A short narrative description.

In this way, we state what the actual requirement is and its origin, and then explain to the person who needs to implement this why it is there.

A note on language and style: In writing requirements, just like in the writing of other scientific documents, one should try to keep the language clear and concise and avoid any unnecessary complexity. Please do not fall to the temptation of trying to make the text "more interesting" by using synonyms. As an example, consider the two statements below:

• If loading the image fails, the user shall be alerted.
• If saving the image fails, the user shall be informed.

Here, the writer probably used "informed" in the second case as a synonym for "alerted" to make the document read a little better. However, to a software engineer this creates ambiguity. Should "alerted" be implemented differently than "informed"? This is particularly important if the software designer's first language is not the one in which the document is written. The possibilities for confusion are endless.

Our advice is that in a specification document, one should use the same exact word throughout the document to signify the same meaning. These types of documents are complicated enough in and of themselves, and one should not introduce additional complexity by using complex vocabulary.

12.5   Risk Management and Requirements

In writing the system requirements document, we have two major steps: (1) identifying the usage scenarios (users, use cases, use environments) and (2) identifying the set of requirements. At each step in the process, we need to evaluate the project for risk (see Chapter 5) as illustrated very nicely in Figure 10.2, which comes from the IEC 62304 standard [119]. Note that in the figure, the "Risk Management" box extends all the way across the software development process. This reinforces the principle that risk (including cybersecurity risk) should be evaluated and managed throughout our project.

At each step, therefore, we must evaluate our project and ask ourselves the following questions:

1. What could go wrong here? (*Harm*)
2. What are the consequences if this happens? (*Severity*)
3. How likely is it that this will happen? (*Probability*)

We then evaluate each of these concerns and try to manage the risk by making changes to our design. This could be done in many ways. Below are some examples.

Sometimes risk enters a project directly as a result of a specified user need. We may need to evaluate this need and see how necessary it is to the project as a whole (and negotiate with our users). Sometimes needs are actually wishes masquerading as needs (i.e. "nice-to-have" as opposed to "got-to-have"). If the need creates a serious level of risk and is more of a wish, one possible solution is to restructure our project to not address this particular item and therefore eliminate the risk by design. This is the ideal solution from a risk management perspective.[4]

Another approach when assessing risk is to add requirements (most likely non-functional requirements) to either design the risk out or to reduce either its severity or probability to acceptable levels. For example, in our image-guided neuro-navigation software, one of the concerns might be that the position of the surgical tools could be out-of-date because of network interfacing issues. There is no way around this as a need: We must be able to show the location of the tool. The solution might be to ensure that, if the last measurement of the position we have is older than (for example) 1–2 seconds, we hide the position from the display or issue a warning to the surgeon in some fashion to make sure he or she knows that what is being displayed is not current information. This can be incorporated as a new non-functional requirement. This is not something the user would necessarily think they needed (what they need is the position of the tool), but rather a way to ensure the safety of the system.

The risk management process should be repeated once the requirements list is finished, to identify additional potential hazards and address these as appropriate. Sometimes the level of risk associated is low/acceptable. This should be noted and documented appropriately.

12.6   Reviewing System Requirements

When reviewing the system requirement document, the following items must be checked:

- Does the statement of user needs, users, and user environment match the expectation of our end users (or customers)?
- Is there anything missing?
- Is there anything superfluous? Given the cost of writing software, we want to write the minimum amount of code.
- Is the origin of each specification a result of one or more use cases?
- Are the needs of each use case completely addressed by the specifications?
- Is each requirement usable as an input to both a software design process and a validation process?

- Do we have quantitative measures everywhere?
- Is any information missing/unclear, particularly with respect to the user environment, that creates ambiguity in the design of the software?

When this document is completed, it must be reviewed and signed off by all the participants (or management) in the process. Then it is entered into the design history file of the project and becomes part of the formal record that will be submitted for regulatory review and clearance. This applies both to the original version and to any subsequent revisions.

12.7 Example: The Image-Guided Neuro-navigation Project

12.7.1 Step 1: Identifying the Usage Scenarios

Let us now come back to our image-guided neuro-navigation project. We pick up the story from Section 11.3. In that section we identified the needs of our users. Now we need to formalize these into use cases, and proceed to identify what functionality our software will need to satisfy these needs. We then explicitly follow the formalism of the game Clue (see Figure 10.3) to begin the definition of our usage scenarios in terms of the following components (questions): Who? What? Where? With what?

Scenario 1

- User: Neurosurgeon.
- Intended case: To navigate during a brain cancer neurosurgery procedure. This means being able to see the position of the surgical pointer on the anatomical MRI images of the patient.
- Intended use environment: In the operating room, using a touchscreen computer mounted at eye level.

Scenario 2

- User: Neurosurgeon.
- Intended case: To be able to see the relative position of the surgical tool to functionally important regions of the brain as identified using functional MRI images.[5]
- Intended use environment: In the operating room, using a touchscreen computer mounted at eye level.

Scenario 3

- User: Neurologist, neurosurgeon, and technologist.
- Intended use: To import functional and structural medical images from the hospital PACS (see Section 3.3.2 for a description of this).
- Intended use environment: In an office using a desktop PC.

Scenario 4

- User: Neurologist and neurosurgeon.
- Intended use: To import surgical plans created in external systems that define the surgical resection area for removing the targeted tumor.
- Intended use environment: In an office using a desktop PC.

There are many other potential uses for our system that we will ignore here for the sake of simplicity. However, looking at the list above, we see two locations (office, operating room) and two uses at each location.

12.7.2 Step 2: Initial Risk Analysis

The next step is to identify the risks inherent in each user need. For usage scenarios 3 and 4 (data import) the risks might be (there are probably others):

1. data (images, surgical plans) is imported incorrectly; and
2. data for the wrong patient is imported accidentally.

Obviously these are absolute needs, so we have to address them. For the first one, we may add a requirement that the patient name is explicitly checked at each step in the process and that the patient name appears on the screen at all times. These would be added as non-functional requirements to our project.

For the second step, a solution might be to create screenshots of critical parts of the plan and the original images in the proceeding software (e.g. the radiology workstation where the images are read and plans created) and import these (perhaps JPEG) images in addition to the actual 3D images/plans. Then, at the start of the process, one could view the screenshots and the 3D display of our system side by side to ensure that the importation of the images was performed correctly.

For scenario 1 we have the potential loss of connection to the external tool tracker device. This has already been discussed in Section 12.5, and potential solutions were presented there.

For scenario 2, a potential problem is that the functional images are continuous images and the definition of functional hot-spots depends on the setting of a threshold. Setting this threshold incorrectly during the procedure might create the illusion of a larger or smaller zone. A solution to this might be to ensure that the images imported have been pre-thresholded and quality controlled by a neurologist, such that only a binary on/off map is available in the operating room. This can become part of the definition of the requirements for addressing this scenario.

12.7.3 Step 3: Identifying the Requirements

This step is the creation of the bridge from the user to the software. The functional requirements are those capabilities of our software that must be present to enable the user to do what she needs to do. Each usage scenario will map to one or more requirements. Sometimes, multiple scenarios result in exactly the same requirement, which is a positive event. Please note that we are designing a system, not a single piece of software. It may be that our solution consists of two separate software executables: (1) the one used in the office environment for importing data; and (2) a piece that is used in the operating room during the procedure.

The non-functional requirements address safety/security and regulatory issues. For example, medical images are shown using "Radiology convention" with the left side of the brain shown on the right of the screen.[6] No user will specify this to a software engineer; it will simply be assumed. This is a *legacy standard* that one needs to become aware of. Other non-functional requirements will come from our risk management process. Others will address the performance and ease of use of the software. Finally, with respect to data privacy: One way to reduce data leakage risk is to design our system in such a way that all patient data is removed from it once the surgical process is completed, and stored only on centralized hospital infrastructure. Our system should also require authentication and use encryption to reduce the risk of unauthorized access.

At the end of the requirements identification, we need to redo our risk analysis for each requirement in turn. Do these functionalities present any sources of harm? How would one address these? For example, one of our requirements that addresses the need from scenario 1 will involve creating a procedure for mapping the position of the patient in the operating room to the actual images. This is a process called registration, and involves a variety of techniques including fiducial markers. A bad registration will introduce errors in the navigation (i.e. the position of the surgical tool will not be accurately displayed on the images), leading to all kinds of problems. A user may specify her accuracy level as part of the needs-finding process (e.g. <2 mm). Part of the risk management process here is, perhaps, to ask the user to physically place the tip of the surgical pointer at known points in the anatomy (e.g. tip of the nose, ear lobes, etc.), and to confirm that the pointer location in the image corresponds accurately to the physical location of the pointer.

12.7.4   Some Examples for Our Project

Intended Users, Intended Uses and Intended Use Environment

Intended users: (A) Neurosurgeon performing surgery; (B) neurologist helping to plan surgery, (C) technologist supporting the process.

Intended uses: (A) Neuro-navigation with anatomical data, (B) neuro-navigation with functional data, (C) image data transfer, (D) surgical plan import.

Intended use environment: (1) *Physical environment and location:* (A) touchscreen PC in the operating room with no mouse or keyboard attached; (B) desktop PC. (2) *Platform:* MS Windows 10. (3) *IT environment:* (C) Our system will need to be able to communicate with the PACS database to import images and annotations. (4) *Legacy:* (D) We need to ensure that images are displayed in the same orientation as in standard radiology workstations.

Functional Requirements

Note: This is not meant to be a complete set! More examples of requirements and a template for the whole document can be found in Section 12.8.

1. Requirement 100

   a. The system shall be able to successfully import images from a PACS server.

   b. This is driven by users B and C for intended use C, which in turn is needed for intended use A.

   c. Comments: The ability to obtain images from the PACS server is a critical component, as these images are needed for neuro-navigation in the clinical workflow.

2. Requirement 200

   a. The system shall be able to align the position of the patient in the physical world (operating room) to preoperative MRI images. This must take less than 2 minutes and have an accuracy better than 2 mm.

   b. This is driven by user A for intended use A.

   c. Comments: Neuro-navigation requires mapping from the physical world to the patient images, otherwise we cannot take the position of the surgical tool in the physical world and overlay it onto an image.

Non-functional Requirements

3. Requirement 500

   a. The system shall be able to notify the user if the current position of the surgical tools is out-of-date, perhaps due to hardware interfacing issues. We define out-of-date as more than 1 second old.[7]

   b. This is a safety concern (arising from our risk analysis step).

   c. Comments: We must ensure that the neurosurgeon is not given false information at any point in the surgery as to the position of the tools relative to the preoperative MRI images of the patient. This could lead to serious errors in the surgical procedure and potentially significant health harm to the patient.

12.8   An Annotated Outline

General advice: Number your sections and subsections and use headings liberally (see below). Also liberally use tables and figures as these add significant clarity to your documents.

A. Introduction

1. Goals and objectives: Summarize in one paragraph the big-picture goals and objectives of the project. Feel free to reference the significance section (e.g. "brain cancer affects x percent of the population (see Section 3.1)"). This should consist of a sentence or two about the disease, a sentence or two about existing technologies, and a sentence or two about what's different about what you are proposing. Keep this short.

2. Statement of scope: Present here a brief description of what the software will do from the user's perspective. Then describe how it will fit into the overall environment (import data from . . ., export data to . . .).

3. Definitions: Define the key terms involved.

4. References: This should include citations to regulatory documents and a short list of other papers (e.g. a review of the disease).

5. Significance: This is the background section of the document that tries to answer the question "Why should I care?" This should consist of numbered subsections covering the disease, the specific clinical problem, the current methods of addressing it and some description of how the proposed solution will advance the current state-of-the art (better, cheaper, faster, etc.).

B. Overall description

This section provides the big-picture description of the software. It aims to capture the user's needs. Use tables here and number all items. The below are all fictional!

1. Intended use(s): This can be a table or a numbered list. List one by one the major (user-facing) uses of the software. For example:

a. Use A: To enable a radiologist to read MRI images.
b. Use B: To enable the radiologist to enter reports.
c. Use C: To enable a referring physician to read the report.

2. Intended user(s): This can also be a table. List the users of your software.

a. User A: Clinical radiologist.
b. User B: Referring urologist.

3. Intended use environments: Where will the software be used and how does this affect the design?

a. Environment A: The radiology reading room. This is where the radiologist will read the images. This is a darker room and large high-quality screens are available. The software needs to interface to the PACS database to obtain images and to the EHR database to store the underlying reports. The images will come in DICOM format and the display of the images needs to conform to standard radiology practice of showing the patient's left on the right of the screen, as our software will be used to replace existing radiology workstations. The software must run on Windows machines as this is what is available in the room. (See Section 3.3 for a discussion of the main hospital databases EHR and PACS.)

b. Environment B: The urologist's phone. The urologist needs to be able to read the report and discuss the results with the patient either in an outpatient setting or while visiting in-patients in the hospital. The requirement here is simply for accessing the text report from Epic. The urologist may have either a reasonably modern iPhone or a reasonably modern high-end Android device with minimal screen resolution of 1920×1080.

C. Functional requirements
1. Requirement 100

a. The system shall be able to download images from a PACS server.
b. This is driven by the combination of user A, use case A, and environment A.
c. Comments: The ability to obtain images from the server is a critical component, as without this our application cannot be integrated into the clinical workflow.

2. Requirement 200

a. The system shall be able to display medical images in standard orientations for radiology review, including the ability to select slices and window the images.
b. This is due to the combination of user A, use case A, and environment A.
c. This is standard practice in clinical radiology.

3. Requirement 300

a. The system shall be able to create and store reports in the hospital's EHR (Epic) database.
b. This is due to user A, use case B, and environment A.
c. This is the critical endpoint of the radiologist's workflow. Once the images are read, the system must be able to seamlessly create and store a report for later use.

4. Requirement 400

a. The system shall be able to read back reports from the EHR database.
b. This arises due to the needs of user B, use case C, and environment B.
c. The urologist must be able to access the reports so as to enable both clinical decision-making and the explanation of the situation to the patient and their family.

D. Non-functional requirements
1. Requirement 1000

a. The system shall preserve patient data confidentiality.
b. This is necessary for compliance with HIPAA and other regulatory processes.
c. This implies that the mobile use case (C) will be stored in an encrypted manner and require user authentication. No such constraint exists for cases A and B as the user will be sitting in a secure environment.

2. Requirement 2000

a. The system shall download and display the images in less than 30 seconds in a usual radiology environment.
b. This will allow radiologists to maintain efficiency in their schedules.
c. This is a performance measure for the overall system.

12.9 Conclusions

The system requirements specification is a bridge between the informal discovery phase of a software project (where the users' needs are identified) and the actual design of the software itself. It is the master document for the project and functions as the input for most other design documents, including the software design document and the validation specification. This document will be read by almost all the stakeholders in the project (e.g. the users, the software designers, the validation team, marketing). Its importance cannot be overestimated. For example, without a proper system specifications document one cannot design a validation strategy.

12.10 Summary

- The system requirements specification is the master document for a software project.
- It provides the necessary background and definitions to allow all stakeholders to understand why this project is important and how it potentially fits into the clinical environment.
- It describes who the users are, how they will use the software, and in what environment (and under what constraints).
- It contains the functional requirements for the project – the functionality the software **must** have to meet the needs of its users.
- It also describes the non-functional requirements for the project to ensure that the software meets the necessary safety, security, and data privacy requirements. These often include implicit needs of the user (e.g. ease of use, performance).

RECOMMENDED READING AND RESOURCES

The authoritative description is the international standard IEC 62304. More example templates for the system requirements document can be found at the website www.projectmanagement.com and in the following book:

D.A. Vogel. *Medical Device Software Verification, Validation, and Compliance*. Artech House, 2011.

SAMPLE ASSIGNMENT

Using the work done in the needs-finding exercise (Chapter 11) as input, create a system requirements document for your project. You may follow the template presented in Section 12.8. This document should be approximately 10 single-spaced pages long.

NOTES

1. Another useful model for these requirements is the Kano model, first described by the Noriaki Kano et al. in 1984 [142] (in Japanese). A description in English can be found in Sauerwein et al. [230].

2. Later, when we create our software design document, the items in this will need to be traceable to system requirements!

3. There is a benefit in keeping the requirement short as this helps in developing the appropriate tests. In this case, we could split the accuracy (95 percent) and performance (less than 25 seconds) into two separate requirements. Should changes be needed in the future, we may be able to limit these to only one of these requirements, such as speed or accuracy.

4. Less is more is a good catchphrase for all medical software. Sometimes, though, less is not good enough! The only completely safe software package is the one that does nothing!

5. One of the authors used to spend a significant amount of time attending epilepsy surgeries. He was shocked to see that sometimes the amount of resected tissue was large (i.e. some people had significant parts of their brains removed). As the senior neurosurgeon at the time commented, "No brain is better than bad brain." The human brain is flexible and function can move (i.e. a different part of the brain takes over the function of the removed part) as part of the recovery process. However, some regions, especially those that directly control motor function and speech, cannot be removed as their functionality cannot be taken over (easily) by other parts of the brain and therefore must be spared during resection.

6. The origin of this, apparently, was to present the physician looking at the image, the same view that they would have were they looking at a patient facing them.

7. Given how slow neurosurgery is, 1 second might be overly stringent here. This is meant to be an illustration.

13 The Software Design Document

INTRODUCTION

This chapter describes the software design document. This is the first step in the software life cycle where we leave the abstract plane and begin to think more concretely about the product. The structure of this chapter is as follows. First, we discuss (Section 13.1) that we are at the point of creating a bridge from the system requirements to the actual code. Next, we present a brief review of key regulatory issues (Section 13.2) involved in creating this document. We then describe a template for the software design document (Section 13.3). As always, we then revisit risk management (Section 13.4) and address how to evaluate our design for potential harms. Within risk management we address the important issue of human factors (Section 13.5). The chapter concludes (Section 13.6) by continuing our discussion of the example image-guided neuro-navigation project by presenting an outline of what the software design document for this might look like.

13.1 A Bridge from Requirements to Code

At this step in the life cycle, the software designer will take the high-level system requirements (see Chapter 12) as the input, and use these to create the actual software architectural design. This document will be used to guide the programmers who will write the code that will form the concrete incarnation of the overall project. It is a bridge between the abstract system requirements and the concrete implementation that is to follow.

13.2 Regulatory Background

In creating our software design document, our primary guidance comes from three documents. The first is the GPSV [70] (Section 5.2.3: Design). The second is the document *Software as Medical Device (SaMD): Application of Quality Management System* [127] (Section 8.2: Design) from the IMDRF. The third is IEC 62304 [119]. The IMDRF [127] summarizes the goal of the process as follows:

The purpose of the design activity is to define the architecture, components, and interfaces of the software system based on user requirements, and any other performance requirements, in line with the intended use of the SaMD and the various clinical and home use environments it is intended to operate in.

This is what we are trying to accomplish, and the recipe presented in this chapter is one way to do this. Next, we move to the GPSV [70]. We quote the opening paragraph of Section 5.2.3 in full, as this is a good description of why having a formal design is necessary:

In the design process, the software requirements specification is translated into a logical and physical representation of the software to be implemented. The software design specification is a description of what the software should do and how it should do it. Due to complexity of the project or to enable persons with varying levels of technical responsibilities to clearly understand design information, the design specification may contain both a high level summary of the design and detailed design information. The completed software design specification constrains the programmer/coder to stay within the intent of the agreed upon requirements and design. A complete software design specification will relieve the programmer from the need to make ad hoc design decisions.

This is really the mission of the work described in this chapter. The key is to take the software requirements (in our vocabulary, system requirements) and create a design that can be implemented by programmers/coders without them having to make ad hoc decisions. It should be complete and detailed so that from here one can simply code!

The next paragraph makes an important point about addressing human factors. The software design needs to be evaluated for ease of use. To quote from the document once again:

Use error caused by designs that are either overly complex or contrary to users' intuitive expectations for operation is one of the most persistent and critical problems encountered by FDA.

One must remember that our software will be used by humans (often operating on little sleep and in a high-stress situation), and as such should be easy and intuitive to use.

The proposed recipe described in the rest of this chapter follows the Software Architectural Design Section of IEC 62304 (Section 5.3). In addition, one may need to create a "software detailed design" (IEC 62304 Sec 5.4). The "detailed design" is not required for Class A software.[1]

Finally, in IEC 62304 [119] we find the acronym SOUP. This stands for software of unknown provenance. This is a piece of software (e.g. library or operating system) that is already developed by others and is generally available (e.g. popular open-source libraries). This might be "off-the-shelf software," or software previously developed for which we do not have adequate records of the process. Again, code is not enough; we need documentation and process. We are responsible for all the components of our software, including those we did not write ourselves. One must identify all SOUPs and justify why they are being used and describe what steps are taken to ensure that they behave correctly – see also Section 13.3.5.

Many medical software packages rely on open-source libraries. In addition to the aforementioned quality requirements, one must ensure that the license of such a library is compatible with a commercial process. For example, BSD-style licenses are generally fine, as opposed to viral-style GPL licenses (see Rosen [221] for an explanation of these terms). With open-source libraries, one must extract and freeze a version of the library (which is constantly evolving) and then perform the necessary testing and validation with this "frozen" version.

13.3   Writing the Software Design Document

When starting to write a software specification, our first step is to carefully review the system requirements document.[2] The intended use environment sections should be reviewed carefully as they may impose fundamental constraints, such as the hardware the software needs to run on.

Once we understand the external constraints imposed on our system, we then construct a set of workflows that will allow our software to satisfy the system requirements. Each system requirement may need a separate workflow, or it may be that multiple system requirements map to the same exact workflow with minor adjustments. Other requirements (especially non-functional requirements) may impose constraints as to how one or more workflows are designed (i.e. they may define how a particular piece of functionality is implemented as opposed to what is implemented). See Figure 13.4 for a concrete definition of a workflow.

The workflows will help us identify the functionality needed by our system (and particularly shared functionality among the workflows). Using these, we then proceed to create a schematic of the software architecture – see Figure 13.5. This will specify the overall structure of our "solution" that is divided into a set of clearly defined modules. We may design a solution that uses more than one piece of software, as in the case shown in the figure, or as is common in many client–server setups.

After this, we begin to design the core user interface as outline sketches to help us establish the basic look and feel of the software. Here, again, we should review any legacy environment issues that may constrain the look and feel of the software, as the users may have strong expectations of what this will look like. Once these steps are completed, we can begin to fill in the details of the workflows and the user interface (see Figure 13.6 for an example).

Once the core functionality of the software is somewhat finalized, we should examine whether we will be using any external libraries for some of the tasks and whether these have secondary dependencies. We need to be careful here to ensure that we only use external code that is reputable and widely used – see Section 6 of the GPSV [70] document for some more information on this topic – and that these libraries are licensed in a way that does not conflict with other product requirements.

Please note that all of these activities are performed before writing a single line of code. One must resist the urge common in all programmers to "hack" a solution and assume that the "paperwork" will take care of itself. The task of creating medical software is not hacking, but more akin to a serious engineering project such as building a bridge. Our code will be used to potentially treat real people, and errors in our software may lead to serious consequences for them. We must take our task seriously and realize that the formal processes in place are there to help mitigate potential problems as opposed to annoy the coder!

13.3.1 An Example Outline

The following outline is a simplified example of how to structure the software design document (a more complex version can be found in Vogel [261]):

1. Cover and revision control pages (see Figure 12.1).
2. Overview – summary of system:
   a. what are the inputs and outputs of the system?
   b. where does it fit in the overall pipeline?
3. Big-picture design:
   a. workflow diagrams (flowcharts) – link workflows to system specifications explicitly (point to specific requirements). This will identify the functionality needed for our system – see Figure 13.4 for an example
   b. module architecture defining the major components of the software – see Figure 13.5, including a description of the modules
   c. outline sketch of the user interface (main window and any views).
4. Specifications of workflows (item by item):
   a. what – background information
   b. how – the actual procedure
   c. how to test – how to test that it works
   d. what could go wrong here (and how to address it).
5. External dependencies – for each one, list: (a) description/author, (b) license, (c) version number, and (d) secondary dependencies.
6. Packaging, labeling, and distribution.[3]

13.3.2 The Overview Section

Medical software often forms part of a user's overall workflow.[4] As such, it will most likely receive inputs from a previous step in the process (and/or a database) and store its outputs in a manner that enables the next step in the workflow to be performed.

    The first critical step, therefore, in designing our software is to understand precisely what comes before and what will follow after our part of the user's "overall" workflow is completed. Hence, the first step in specifying our software is to explicitly describe its interactions with the rest of the "software" universe that our user interacts with. This includes a detailed description of file formats, database/network protocols, and all other issues involved here.

13.3.3   The Big-Picture Design Section

Workflow(s): This step consists of the design of the process that is to perform a particular task/satisfy a particular requirement. We present an example at the end of Section 13.6 for the patient–image registration module. The workflows are the functional steps that are needed for our application to meet the system requirements. In a complex application, we may need to subsequently break each of these boxes into its own workflows. These should be specified from the perspective of the user (as much as possible).

If our software involves the use of an artificial intelligence/machine learning (AI/ML) model, we should also include a workflow for the training of the model and the management of the appropriate training data (see Section 9.3) even though this functionality will not be typically included in the software supplied to the user at the end of the process.

Architecture design: Here we describe the overall system, its major subsystems and modules, and the communication links between them. For our neuro-navigation example, we might start with an overall diagram such as the one shown in Figure 13.5. In this particular case, our system consists of three subsystems that communicate using network connections. It also communicates with the hospital database (over a network connection) and with the external tool tracking device (over direct USB connection). This subsystems will allow the software to efficiently implement the functionality required by the workflows.

When designing the software architecture, it is important to ensure that the system is as modular as possible and that, in particular, access to external resources such as databases and other devices is restricted to a single module for each type of connection. This will also allow for easier testing as we create simulator software to interface to these "connector" modules to "mimic" the behavior of this external resource. An additional benefit of this type of design is that it makes upgrading the software to use a different external resource, should one be needed in the future, easier, as all the functionality is in a single module that can be replaced without affecting the rest of the software.

Sketch of user interface: This is useful as it constrains what the software will look like prior to the coder taking over – we discuss the process of user interface evaluation in more detail in Section 13.5.

13.3.4   Specifications of Workflow Items

To illustrate this process, we present here an example image-smoothing workflow. This is a simple procedure that consists of five steps:

A. load image;
B. set smoothing level;
C. smooth image;
D. review smoothing;
E. save image.

The specification of steps A and B of this are given below. It follows the usual pattern of explaining why this step is necessary, followed by a detailed description of how to accomplish the task. At the end, we specify testing criteria and anything else that we need to watch out for. The testing criteria are useful for software verification, which is the process of ensuring that software was correctly implemented. This will be used by the software testing team to test the software. If the software cannot be tested, then it is impossible to have any degree of confidence that it was correctly implemented.

Specification of step A: load image

1. Background: We allow loading of JPEG (.jpg) and PNG (.png) files.
2. User clicks on the "Load Image" button.
3. A dialog box pops up instructing the user to select an image having a suffix .png/.jpg.
4. If a file is selected, the image will be loaded appropriately.
5. If the image loading fails, we output a warning to the user and stop.
6. Else we will proceed to Step B.
7. To test: load a known image and compare to the snapshot from the previous working version.
8. What could go wrong? If the image is above a certain size we could run out of memory. Check that the image size is less than a selected maximum and alert the user if it is too big.

Specification of step B: set amount of smoothing

1. Background: The amount of image smoothing is the standard deviation of the Gaussian filter that will be used in Step C to smooth the image. This needs to be set by the user. The default value is 2.0 as it is appropriate for this type of image.
2. A pop-up dialog box asks the user to set the amount of smoothing in the allowed range (0.5–10.0 voxels).
3. If the user presses "OK" then proceed to Step C.
4. If the user presses "Cancel" then close the dialog and wait for user input.
5. To test: If the users presses "OK," does smoothing happen? If the user presses "Cancel," does it stop? Is smoothing set staying in the allowed range?
6. What could go wrong? If the image is too large the process could take a significant amount of time, and/or result in a system crash. This should be addressed by the previous step.

13.3.5   External Dependencies

In this section, we describe the external libraries and tools that will be used to implement, compile, and test our software. We should be explicit in stating where these tools are obtained and what version of the tool we are using, and offer some justification as to why this tool was selected.

Finally, one should (within reason) try to minimize the dependency of the software on external libraries ("SOUPs" [119]). This is code that we have no control over and which may introduce additional risks that we need to manage. Naturally, one should not constantly reinvent the wheel. The best compromise is to use well-known, widely used, and well-tested libraries that have a reputation for quality and that have been used by other important projects. While we do not have the luxury of expecting that such code (especially free, open-source libraries) will have been developed within a full quality management system (QMS), we can always look for libraries whose development process has some of these attributes (e.g. code reviews, detailed testing).

13.3.6   Packaging, Labeling, and Distribution

The software should be distributed with an appropriate set of labels that describe the origin, copyright holder, regulatory compliance level, what the software is cleared for, and any other information that is necessary for its use – see Figure 13.1 for an example.[5] In the medical world, software is often installed at a user site in combination with hardware (e.g. an image-processing workstation), so this step may involve integration with other parts of the underlying business unit that oversees the overall product.

Figure 13.1 Labeling guidelines for apps (UK MHRA). This figure illustrates how regulated mobile apps should be labeled. The guidance states that a "prospective buyer should be able to identify that the app meets the relevant essential requirements prior to purchase. As such, a developer should display the UKCA mark on the primary landing page and as a screen shot in any app store." The term "UKCA" stands for UK Conformity Assessment, and is the indication that this software is cleared for use in the UK. Reproduced from the MHRA, the UK regulatory authority [174].

13.4   Risk Management

13.4.1   Overall Process

In specifying the workflows (see Section 13.3.4), we already include the question (for each step) "What could go wrong?" This is effectively risk analysis at the low level. However, one needs to review the overall structure of the design for risks as well. Is the user interface confusing? Could it lead to misuse? Would failure at any part of the workflow(s) create a hazardous situation for either patients or users? Is there complexity risk in some of the individual steps such that (1) it could result in buggy code or (2) it may not be feasible to implement, given our resources and the levels of technical and scientific expertise in our team? Will a particular step be costly to implement to a degree that will endanger the whole project?

These and more questions need to be addressed to ensure that we maintain as low a level of risk as possible. It may be that the risk analysis of the software design necessitates changes upstream to our system design, such that we have to stop this step and move back to redesigning the system. The risk involved in the implementation may, in this case, not be addressable/manageable simply by better software engineering techniques. It could be caused by an over-optimistic high-level system design. This is why, in the review process for the system requirement, one should include the software engineers who will actually perform the detailed design of the software, as they might be the ones who have the expertise to point out some of these issues. Experience may also dictate that certain components are likely to be high-risk due to complexity and/or a history of having issues in the past. We will revisit this topic in Section 14.2.3.

13.4.2   Software Design and Cybersecurity

We should also evaluate whether our design results in any particular cybersecurity issues. This is particularly important if (1) we are storing patient data, or (2) our software provides "server" functionality over the network. Issues include authentication (for the users and remote services) and encryption for data (both when stored and when transmitted) – see Section 3.5 for more details. We should also evaluate any dependencies for known cybersecurity risks (vulnerabilities) and ensure that appropriate mitigation measures are in place. The reader is referred to Wirth [269] for a comprehensive coverage of this important topic.

13.4.3   Human Factors and Software Design

The design risk management process involves a consideration of human factors. Our discussion here is based primarily on the FDA's guidance document [89]. A good starting point in the discussion of human factors is the model of device–human user interaction shown in Figure 13.2, reproduced from this document.[6] This shows the cyclical flow of action and information between the user and the software. We can start the cycle at some software output which is perceived and then analyzed by the user.

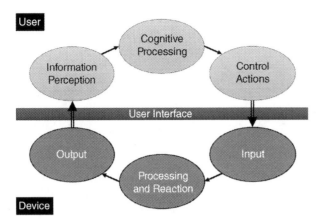

Figure 13.2 Device User Interface in Operational Context. This figure is replicated from the FDA's guidance document on human factors [89] (figure 4 – this was adapted, in turn, from Redmill and Rajan [213]).

This analysis then leads to some form of user action (through the user interface of the software). This action acts as an input to the software which performs some processing, resulting in new software output (that appears on the user interface) that starts the next interaction cycle.[7]

In performing a human factors analysis, a first step is to identify known use-related problems in similar software. This involves the usual mixture of experience with previous products and potentially obtaining information from existing databases, a number of which are listed in the guidance [89]. The other steps may involve having experts in human factors engineering review a user interface against standard guidelines to identify potential issues. In addition, domain experts (e.g. radiologists) should check for potential hazards based on their experience with similar systems.

Section 5.1 of IEC 62366-2 describes how features that improve safety may also improve usability. For example, placing limits on parameter settings (e.g. using a slider as opposed to a text entry box) both improves the safety (i.e. the user is unable to set unsafe values) and also improves the usability (i.e. it reminds users of the allowable range) as it reduces the cognitive load on the user and increases confidence in the software. Similarly improving the speed at which a user can perform a task (usability) can have an effect on safety. For example, if the software interface slows down the user it can lead to them skipping over some steps (some of which are critical safety checks), resulting in reduced safety of a given procedure.

A human factors analysis will also most likely provide information on how to improve the user experience, thus making the software easier and more pleasant to use – see also Section 6.2. IEC 62366-2 [124] has an annex titled "Developing usability goals for commercial success" that is worth reading for more information on this topic. In the next section, we will explicitly discuss user interface design and evaluation based on the description in IEC 62366-2.

13.5   User Interface Evaluation

IEC 62366 defines two types of user interface evaluation:

Formative evaluation: A user interface evaluation conducted with the intent to explore user interface design strengths, weaknesses, and unanticipated use errors.

Summative evaluation: A user interface evaluation conducted at the end of the user interface development with the intent to obtain objective evidence that the user interface can be used safely.

The first type of evaluation (formative) is performed during the actual design of the software, whereas the second (summative) is performed at the end of the process to confirm that the user interface is safe. The standard [124] provides an outline of the process, which is summarized in Figure 13.3. We discuss each step of the process here:

1. **Prepare use specification:** This builds on our discussion in Section 11.2. The goal here is to identify the users, their environments, and the medical situation in which our software will be used. This will involve a variety of techniques such as direct observation, interviews and surveys, expert reviews, convening an advisory panel, and performing usability tests on existing/comparable devices (if they exist).

2. **Perform safety analysis:** This is fundamentally a risk management process. The first step is to evaluate user interface characteristics that pose a threat to safety and identify hazards and hazardous situations. Next, we identify use scenarios in which hazards may occur and select the most critical of these (e.g. based on severity) for summative evaluation – see Step 5.

3. **Design:** Here we specify the user interface (i.e. what the required functionality is) and the accompanying documentation and training materials. As the standard states, "Because the instructions for use and other accompanying documentation are considered part of the medical device, user interface requirements should be developed for instructions for use and other accompanying documentation as part of the user interface specification."

4. **Formative evaluation:** This is an iterative process. In a first iteration, we may develop design concepts and then create a wireframe-type prototype (using tools such as Figma [93]) that is evaluated simply by doing a walkthrough. In subsequent iterations, the user interface progressively acquires higher fidelity and perhaps we initially evaluate this using expert reviews (e.g. asking radiologists to evaluate the

Figure 13.3   User interface design and evaluation. This flowchart summarizes the process of user design and evaluation as described in IEC-62366 [122, 124].

interface for an image viewer). Eventually this is refined to the point where we implement it in a version of the software and perform a formal usability evaluation (measuring both performance and errors and collecting subjects from test users). This step may be repeated a number of times until the performance measures reach acceptable levels.

5. **Summative evaluation:** This is performed at the end of the process, with the final software, and is part of the validation of the software. The goal is to demonstrate that the device/software is not vulnerable to potential use errors. Again, we collect both performance data and general comments. The performance data will include both (1) correctness categorized as success, close call, and error; and (2) an evaluation of the difficulty in performing a given task. The comments will allow us to obtain an overall impression of the software and instances of confusion/difficulty.

13.6 Example: The Image-Guided Neuro-navigation Project

We continue here the story from Section 12.7. What does our system look like? What is the user interface that will be presented to the user? What are the workflows that we need to implement?

13.6.1 Workflows

Our system will need to implement functionality for multiple workflows to satisfy the needs of our users. This is an incomplete list:

1. import structural data from a clinical database, and verify and store it for use during surgery;
2. import surgical plans from the clinical database, verify quality and store them for use during surgery;
3. import functional MRI data from the database, threshold to the appropriate level and store the binary functional "hot-spots" for later use during surgery;
4. perform patient–image registration (described in detail below);
5. display anatomical and functional images and tool location.

The workflow for the patient–image image registration is shown in more detail in Figure 13.4. Please note that the last three steps in this are essentially risk mitigation steps that satisfy non-functional requirements identified during risk analysis. In the full document, we would include flowcharts for each of these workflows.

13.6.2 Architecture

Once the workflows are specified, we use these as inputs to design the overall solution. One possible design is a system that will consist of two distinct pieces of software.

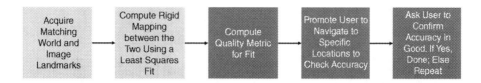

Figure 13.4 The patient–image registration workflow. One of the workflows for our software is the patient–image registration workflow. This has five steps. The first two steps (shown in light gray boxes) are required to satisfy the stated needs of our users (i.e. they come from the functional requirements). The last three steps (shown in darker gray boxes) originate from non-functional requirements that were arrived at after a risk analysis process.

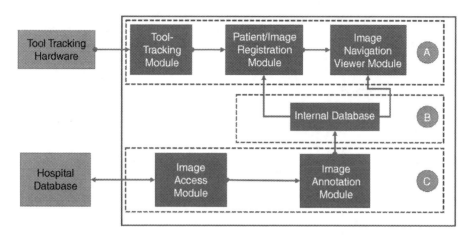

Figure 13.5 The overall system design for our neuro-navigation example. Our software system consists of three pieces of software: (A) the neuro-navigation module for use in the operating room; (B) our internal database server to store information about procedures; and (C) the desktop module for use during the planning phase to import images and other information. The external links are to the main hospital database and the tool-tracking hardware. Subsystems A and C are, in turn, subdivided into major modules as well.

The first is the package that will be used outside the operating room for importing images and ensuring that everything is ready for the surgery. The second piece of software is what the neurosurgeons will use during the actual procedure. This solution has significant advantages as we can have a more traditional user interface for the first system (where a mouse and keyboard will be available), thus simplifying some of the operations there, whereas we will be restricted to a keyboard-less/mouse-less touchscreen-only setup in the operating room. A further review introduces the need for a simple, internal database server to store data as we move between the two phases.[8] Hence, we arrive at the three-subsystem solution shown in Figure 13.5.

One should then describe the functionality in each of these modules and how this satisfies the needs of the workflows above. For example, we would specify that the *image access module* will contain functionality to interface to the hospital image database to obtain preoperative image, for the patient. This will be involved, for example, in the first three workflows specified in the previous section.

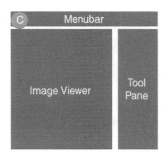 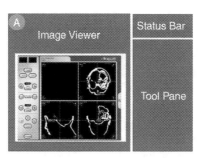

Figure 13.6 The user interface schematic for our neuro-navigation example. The left view is the UI for the desktop module (C in Figure 13.5). The right shows the UI schematic for the neuro-navigation module for the touchscreen setup in the operating room (A in Figure 13.5). The picture inside the image viewer block for subsystem A shows the user interface of the BrainLAB VectorVision Cranial System from around 2005, when "operating" on a rubber-head phantom.

13.6.3   User Interface

We show a schematic of the user interface in Figure 13.6. Depending on how critical the individual elements are, this type of figure will then be augmented by additional figures showing, for example, sub-views and dialog boxes.

13.7   Conclusions

The software design document is a bridge between the system requirements specification and the actual process of implementing the software (coding). It is the master document for the implementation process. This is typically written by the senior software engineers in a team as a guide to the (often) more junior software developers (programmers) that will implement the software. The goal of this document is to "relieve the programmer from the need to make ad hoc design decisions" [70]. It should include enough detail that the programmer can focus on the job of figuring out "how" the software should be implemented without being concerned about "what" the software should be doing.

13.8   Summary

- The software design document is the master document for the software implementation process.
- It describes both the high-level software architecture and the user interface of the proposed software.
- It should provide enough detail to ensure that critical decisions are not made at the time of coding.
- It should include a list of all external libraries/resources that the software will use.

- All software modules should derive explicitly from the functional and non-functional requirements described in the systems specifications document.

RECOMMENDED READING AND RESOURCES

There are many prototyping tools that can be used to create mock-ups of the user interface. One such tool is:

Figma. https://www.figma.com.

SAMPLE ASSIGNMENT

Based on your system specification document (the assignment for Chapter 12), create a software design document for your project. The document should be approximately 5–7 single-spaced pages long. Please submit a draft of this document a week prior to the deadline for peer review.

NOTES

1. We discuss classifications in Section 10.4. Software that is classified as belonging to classes B or C will need additional controls and more detailed design, as detailed in IEC 62304 [2, 119].
2. This naturally must exist and be completed before we can start this part of the process. Otherwise we are designing software in an out-of-control manner!
3. For class assignment purposes, this can be very brief.
4. As was discussed in the description of the specifications document in Chapter 12, in a more formal setting there will also be an additional section (or two) prior to this overview section that will provide a set of definitions and references.
5. Strictly speaking, the label should list the intended use and indications for use – see Chapter 2.
6. This is known as the PCA model, where PCA stands for perception, cognition, action. Some readers may also be familiar with the statistical technique PCA, where PCA stands for principal components analysis. The take-home lesson is that when faced with an acronym, one must ensure they have the correct meaning!
7. This model can be used by a software designer to customize the functions and features to best meet the user's needs. Overall, this leads to a tight relationship between the medical software designer and the user of the product, which is often the key to a successful implementation of the software.
8. This is in fact how the BrainLAB system functioned in the early 2000s. However, that system did not use a database server but instead stored the data on a physical disk to enable data transfer from one system to the other. Storing patient data on portable disks would now be considered a major data privacy risk, as these are easily misplaced or stolen. We may also be able to use an existing hospital database solution for this task.

14 Software Construction, Testing, and Verification

INTRODUCTION

This chapter describes software construction and testing. These are the very concrete steps in the software life cycle (Section 14.1). Next, we discuss "construction," or coding (Section 14.2), beginning with a brief review of key regulatory issues. We follow this with a presentation of programming topics and conclude this section with an extended discussion of risk management in the context of the programming process. The next section (Section 14.3) focuses on testing and again begins with a review of regulatory issues before moving to a brief section on risk management in the context of testing. In the final section (Section 14.4), we provide some pointers as to how one would go about creating a verification plan.

14.1 Two Concrete Steps of the Software Life Cycle

Coding and testing are some of the most concrete parts of the process. This is where the "rubber meets the road" and our ideas need to be turned into an actual product. As a reminder, this *is not a programming book*. We will not discuss how one should write code, other than for some pointers that relate specifically to our topic. Our remarks on coding (or "construction" [70]) mostly focus on the management of the process. The same logic applies to our discussion of software testing. We provide an introduction to this topic in Section 9.5. Our focus here will be on the particular issues that arise in medical software.[1]

14.2 Software Construction

14.2.1 Regulatory Background

The IMDRF QMS document [127] provides useful guidance concerning software construction. This extract (from Section 8.3) is a good summary:

The development activity transforms the requirements, architecture, design (including interface definition), recognized coding practices (secure), and architecture patterns into software items and the integration of those software items into a SaMD.

The result is a software item/system/product that satisfies specified requirements, architecture, and design. Good development practice incorporates appropriate review activities, (e.g. code review, peer review, creator self-review) and follows a defined implementation strategy (e.g. build new, acquire new, re-use of existing elements). Design changes resulting from the review activity or development activity should be adequately captured and communicated to ensure that other development and QMS activities remain current.

Use of appropriately qualified automated tools and supporting infrastructure is important for managing configuration and having traceability to other life cycle activities.

Again the focus here is on the management process and not on the actual programming. The most interesting comment has to do with static code reviews – a process also recommended in the GPSV (Section 5.2.4) [70]. This involves experienced programmers reviewing the code statically (i.e. reading the code without executing it) and looking for logical flaws and poor structure – see Section 9.6.2 for an example on the use of GitHub to accomplish this.

14.2.2 Comments on Programming

Our recommendation is to follow the usual best practices that one is taught in most computer science settings. Code should be appropriately modular, well commented, and easy to read.[2] Defensive programming techniques should be used. At the beginning of the project, one should establish a common repository and changes should be tracked and reviewed appropriately – see Section 9.6.1 for more information on this topic.

One unique aspect of medical software is its longevity. Many medical software packages are used and maintained for extended periods of time (decades). For this reason, the code is highly likely to be reviewed and modified by programmers that were not present[3] at the time of the design. This places an extra burden to write very clear and easy-to-read code and to avoid being overly "smart" and "terse" as programmers are wont to do. One should optimize for readability over compactness, and perhaps even allow for some reduction in the speed and efficiency of the software if this improves the readability of the code, so long as it does not compromise the satisfaction of the requirements.

Inevitably, medical software will end up being used in different countries with different language requirements. It is important, therefore, to design the software with internationalization in mind. This will enable switching all messages/user interface elements from, for example, English to Japanese without needing to perform major surgery to the code. The usual recommendation is to keep all strings in separate tables/files that can be easily translated and substituted depending on the user's environment.

A final comment regarding external libraries. The external libraries (or SOUPs – software of unknown provenance) for our project are specified in the software design document. If during the code implementation phase we discover that we need to use an additional library, the correct procedure is to stop coding, ask that the software specification be revised to include this, get the revision approved, and then – and only then – continue programming. Developers should not add dependencies without prior approval.[4]

14.2.3   Risk Management

We return to our constant companion, risk management. At this point in the process, we have already evaluated the requirements of our users (and associated risks) and have come up with a software design (and assessed the risks involved in that). So theoretically all that is left to do is to write the code for a workflow that has already been assessed and evaluated for risk, and which has been modified to account for this. What could possibly go wrong?

One of the core truths about engineering (and software engineering is just that) is that the real world gets a vote. We are now leaving the realm of plans on paper and moving to actual code. Our plans are about to be tested by contact with reality. As the Prussian Field Marshall Helmuth von Moltke is reputed to have said "No plan survives contact with the enemy!"[5]

Hence, despite the best efforts of our strategists/generals (such as designers and business analysts), it is now down to the foot soldiers (programmers) to execute this plan. All code written by humans is prone to errors, and our risk management plan here essentially consists of identifying those parts of the code that are likely to fail (due to human error) and to come up with mitigation strategies for controlling for the risk created by such errors.

As a reminder, risk is a combination of probability and severity. Hence, in evaluating our code for risk, we need to estimate both the probability that something goes wrong and the severity of the harm caused by such an event. In some ways, severity is easier to identify and ameliorate, as we can examine what effects failure in any part of the code could cause and come up with an objective estimate.

Probability, on the other hand, is harder. Software is not hardware. Software is deterministic, and as such failure occurs when the software encounters conditions that cause it to fail.[6] Therefore, we need to use experience and intuition to identify pieces of code that are likely to fail and attempt to create solutions for that.

Vogel [261] presents a useful checklist of the types of things that increase the probability of failure for software. Some of these are (1) large size and great complexity of the particular item, (2) lack of skill and/or experience of the developer, (3) any external stress in the circumstances in which the code was created, such as deadline pressure and late changes in the specifications, (4) aspects of the code that have a history of failure, and (5) issues with testing of this particular part of the code, such as limited or incomplete testing or testing performed by inexperienced testers.

To mitigate such risk, we have a number of strategies available. The first (and obvious) one is to assign tasks based on the skill and experience of the developers. The second is, in the case of complex/large items, to revisit the design to simplify either the size or the complexity of the items and to perhaps create a more detailed design prior to coding for such items. A third option is to ensure that parts of the code that carry significant risk (either severity or probability) are tested thoroughly under a variety of circumstances.

Finally, sometimes we need to acknowledge that while the design was a good idea, it cannot be safely implemented, and then go back to the drawing board to change the upstream documentation (software design and/or system requirements) to account for that.[7]

14.3 Software Testing and Verification

14.3.1 Regulatory Background on Software Testing

We presented introductory information on software testing in Section 9.5. Unlike coding, the regulatory documents dedicate significant space to testing. The general description is divided neatly between testing done by the developer (GPSV Section 5.2.5) and testing done at the user site (GPSV Section 5.2.6). IEC 62304 [119] divides the process into three steps: (1) unit testing (Section 5.5); (ii) integration testing (Section 5.6); and system testing (Section 5.7). In our discussion of software testing, we will follow this second division, though the two are in fact very similar. We presented background material on these levels of testing in Section 9.5.3.

The GPSV has a useful checklist that explains how the testing is targeted. Using this we arrive at the following guidelines (paraphrased):

- Unit (module) tests are traced to the detailed design (needed for classes B and C).
- Integration tests are traced to high-level software design – this is what we described in Chapter 13.
- System tests are traced to the original software requirements – see Chapter 12.

This is analogous to the V-model in Figure 1.3. The statements above emphasize again the importance of documentation. The testing cannot be performed in the absence of these documents, as the design of the tests begins from these.

We note that unit testing and integration testing are only required for software classified in classes B and C (regarding harm). System testing is now required for class A software as well – this was not required in the 2006 edition of IEC 62304, but was added as part of the 2016 amendment [2].

The IMDRF QMS document [127] (Section 8.4) provides some additional guidance. We excerpt here two quotes:

These V&V activities should include scenarios that cover the clinical user/use environment (usability, instructions for use, etc.). This can be accomplished, in part, through structured human factors testing using a subset of patients/clinicians.

Interoperability of components and compatibility to other platforms/devices/interfaces etc., with which SaMD works should be considered.

The first quote reminds us that the testing needs to include testing that mimics (or uses) the environment of the actual user. Many issues are quickly manifested when we place our software in this real-world situation ("no plan survives contact ..."). The second statement is a reminder that our software is almost never going to be an island. It will need to interface with other software/devices and therefore our testing should explicitly include this interfacing. This is a different level of integration testing, in which we test how our software is in fact a module in the user's workflow and therefore its integration needs to be tested.

14.3.2 Risk Management

One would think that we are done with risk management. We are, after all, in the testing phase, which is how we ensure that everything works. Why are we still talking about it? The short answer is that our testing may fail to reveal errors/problems with our software. To go back to our earlier analogy of comparing software testing to cancer biopsy: While we may be diligent in terms of looking for cancer, we may still miss it. This "miss" may be a result of human error or because the cancer did not manifest itself under the conditions of the biopsy or perhaps we missed a small metastasis.

Testing, especially at the lower levels such as unit and integration testing, involves the writing of code to exercise the units/software to create the test cases. This is code just like the code we write for our software. As such, this code may also have errors in it that hide the errors it is designed to catch.[8] Hence, testing code needs to be designed and reviewed with similar rigor to that used for the production code.[9]

Testing may also fail because we do not expose the software to enough external conditions in which failures might manifest themselves. If the software is always tested by the same person, in exactly the same way, this may yield false confidence that the software works, as one might fix any issues that arise in this very limited scenario.

When evaluating risk in testing, we again return to severity and probability. Estimating probability of failure (i.e. that the *testing code/process* fails) is, in many ways, similar to the process of estimating the probability of failure of the actual software, as described in Section 13.4. To summarize: inexperienced testers operating under difficult conditions with ill-defined tasks are likely to make mistakes.

Please realize that it is impossible to completely test all aspects of anything but the most trivial piece of software. The key is to reduce risk to an acceptable level – see Chapter 5. In an early article, Littlewood and Strigini, after concluding that exhaustive mathematical proofs of software correctness are impossible on real-world systems, advise that it may be best:

simply to accept the current limitations of software and live with a more modest overall system safety. After all, society sometimes demands extremely high safety for what may be irrational reasons. Medical systems are a good example. Surgeons are known to have fairly high failure rates, and it would seem natural to accept a computerized alternative if the device is shown to be as good as or only slightly better than the human physician. [158][10]

14.4   Creating a Testing and Verification Plan

We will not discuss in detail the creation of the verification plan. Vogel [261] is an excellent resource on this topic and the interested reader is strongly encouraged to read this in depth. In general, though, for class A software (i.e. low risk), the plan should in outline form proceed as follows:

1. For each requirement, design a set of tests to demonstrate that it is met. For example, if our requirement is that the system is able to obtain images from a database server, we have a test in which a tester is asked to download specific images from a server and to confirm that these images are downloaded correctly.
2. Create tests that test the entire workflow, end to end. Here, a user is asked to mimic a real-world procedure situation. In many of these, one may need to create controlled conditions (e.g. a rubber-head phantom for neurosurgery – see also the example shown in Figure 14.1) to allow one to test in pseudo-realistic settings.

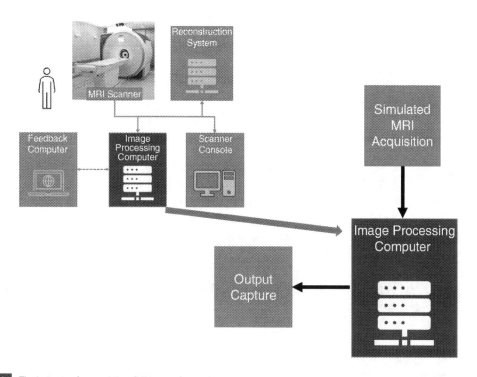

Figure 14.1   The test setup for a real-time fMRI neuro-feedback system [105, 260]. In real-time functional MRI, the subject is given feedback as to what the level of activity in specific regions of the brain is. He or she is then asked to try different strategies to change that level. It is a potential treatment technique for many psychiatric disorders. The system involves three components: (1) the MRI scanner, (2) the feedback computer, and (3) software to process the data in real time. This is shown in the left side of the figure. Our "system" was the "image processing computer." The rest are external devices. To test this, we had to create two simulators: one to create simulated MRI data and one to capture the output of the system. Simulators are a powerful tool for developing and testing software that needs to interface with external hardware or other external systems. The picture of the MRI scanner is from the website of the department of Radiology and Biomedical Imaging, Yale School of Medicine.

These tests may involve replicating previous measurements and ensuring that our accuracy/efficiency (or any other appropriate measures) remains acceptable or similar to previous cases.

3. Add a set of random operations (e.g. ask the tester to "play" with the software) to see if anything breaks in such unstructured/informal testing.

Each of the items described above needs to be explicitly "written out" with clear steps for the tester to follow and explicit instructions as to what they need to report at each step of the process. A good description will include plenty of pictures (screenshots of the software at different stages of its operation). There also needs to be a process for reporting the results and reviewing them with the development team and project leadership/management.

Regression testing for automated parts of the processing should also be added at the system level. Ideally, one should be able to execute the same executable that the user uses in "testing mode" in which all inputs are provided as part of a command line interface, and obtain outputs that can be compared to previous "silver standard" results.[11]

This is not a complete recipe by any means. Naturally, for example, we also need to check not only for correct functionality (verification) but also whether the users like the software and enjoy using it. Some of this "desirability" can be formalized through extra non-functional requirements that explicitly try to make the experience enjoyable. Alternatively, one can use a more informal process by obtaining feedback from early beta testers and other customers. While user delight is not a regulatory requirement, it can be a market requirement, especially if there are competing products. A perfectly functional but clunky software will not sell well, and all the other effort and investment in its design, development, and regulatory approval will be wasted.

All of these processes should be collected in a testing/verification manual that should be kept in the same design history file as all the other documents described so far (e.g. system requirements and software design documents).

For those systems involving hardware/software combinations, software that simulates the functionality of the hardware can be extremely important in enabling both the development and the testing process. The same applies to tools that simulate external databases such as EHR and PACS – these are discussed in Section 3.3.

14.5   Conclusions

This chapter covers the most concrete, low-level, aspects of the medical software life cycle. The software must be implemented and the code tested to make sure that it works (verification) before trying to establish that it meets the user's needs (validation). Both the coding/construction aspect of this process and the testing component follow all the usual best practices one encounters in computer science classes. The one additional issue is that medical software tends to have a long lifetime and that one must explicitly write code in such a way as to be easy to read and modify by others later. Finally, it is worth remembering that it is impossible to completely test anything other than the most trivial piece of software. The goal is to reduce risk to an acceptable level by a combination of proper design, high-quality implementation, and appropriate safeguards.

14.6   Summary

- The regulatory documents have little specific to say about software construction or coding. This is because, for the most part, all the usual best practices from other areas of computer programming also apply here.
- An important exception to this is the restriction on the uncontrolled use of external libraries (or SOUPs [119]). Programmers should not add extra dependencies to the software without approval and review (and inclusion in the software design document).
- The process of software testing and verification is also similar to what one encounters in regular software engineering. The regulatory documents simply clarify what level of testing is required for software depending on its risk classification.
- An important aspect of testing is interactive user testing.
- Risk management is an ongoing activity for both coding (how to account for "bad" code) and testing (e.g. incomplete or poorly done testing).

RECOMMENDED READING AND RESOURCES

This chapter directly builds on material presented in Chapter 9.

For a regulatory perspective on programming and testing, see the FDA's classic document:

FDA. General principles of software validation; final guidance for industry and FDA staff, January 11, 2002.

Another useful primary source is the IMDRF document:

IMDRF, SaMD Working Group. Software as a medical device (SaMD): application of quality management system, October 2, 2015.

David A. Vogel's book has an excellent description of software testing and verification issues.

D.A. Vogel. *Medical Device Software Verification, Validation, and Compliance*. Artech House, 2011.

For those curious about real–time fMRI, these are pointers to our previous work:

M. Hampson, D. Scheinost, M. Qiu, et al. Biofeedback of real-time functional magnetic resonance imaging data from the supplementary motor area reduces functional connectivity to subcortical regions. *Brain Connect*, 1(1):91–98, 2011.

D. Scheinost, M. Hampson, M. Qiu, et al. A graphics processing unit accelerated motion correction algorithm and modular system for real-time fMRI. *Neuroinformatics*, 11(3):291–300, 2013.

NOTES

1. It is interesting that the IEC 62304 standard spends significantly more space on the testing aspects than the "code writing" aspects. It simply states (this is the entire content of Section 5.5.1) "The MANUFACTURER shall implement each SOFTWARE UNIT. [Class A, B, C]" [119]. In plain English, it basically says you need to implement your design! That is all.

2. One may also need to explicitly specify that all comments, function names, variable names, etc. are in a specific language (e.g. English). One of the authors once was told of a project (in the United States) where one programmer named everything in Finnish. Once he left, the code was unreadable!

3. One is tempted to say not alive!

4. This is incredibly easy to do, especially in the era of large package repositories such as `pip` (Python) and `npm` (JavaScript).

5. Helmuth von Moltke (1800–1891), head of the Prussian General Staff 1857–1871. Mike Tyson (1966–), the former heavyweight boxing champion, is also quoted as saying something very similar: "Everybody has a plan until they get punched in the mouth."

6. The probability here is the probability of these conditions (i.e. inputs) occurring. The software will fail each and every time such conditions are encountered. It is not like hardware, which, due to either wear and tear or bad materials or failures in manufacturing, has a certain "real" probability of failure.

7. This type of solution is often in evidence when features in a beta version of particular software are removed prior to release. The feature was probably a great idea, but it could not be successfully implemented in the available time.

8. A potential issue that we have encountered was that the test "runner" code declared (erroneously) that all tests passed. This bug appeared in the test reporting phase. One solution to this is to create tests that are designed to fail, to ensure that our test code can spot these!

9. Vogel [261] makes an important comment that if the people who write the testing code never make mistakes, we should ask them to write the real software instead and save ourselves the trouble of writing testing code! Clearly this is not the case.

10. This is a valuable lesson to keep in mind. Quality assurance managers or executives face the common temptation to say "test everything," which makes engineers think to themselves "that sounds impossible or at least like more work than I want to do" and the process breaks down. Both sides need to realize that while exhaustive testing is theoretically impossible, valuable testing is feasible and there are reasonable standards to aim for. This allows for a more productive discussion. As ISO 14971 [133] states, "absolute safety in medical devices is not achievable." The best one can do is to reduce risk to acceptable levels. Appropriate levels of testing are part of this risk management process.

11. An extreme (positive) case is the self-testing module of 3D-Slicer [235]. As stated in the Slicer wiki, the goal of this module is to make the regression test of the software part of the binary distributions, so users can confirm correct behavior on their systems.

15   Software Validation

INTRODUCTION

This chapter describes the process of software validation. We first begin with a brief overview of what validation is. We then review the regulatory guidance for validation (Section 15.1) in general, and then provide an extended discussion on the issues that affect the validation of software modules that use artificial intelligence/machine learning (AI/ML) techniques in particular. Next, we present a description of human subject studies and clinical trials (Section 15.2), including a description of the necessary ethical constraints on such studies. In the last part of this chapter (Section 15.3), we present a simplified strategy for designing a validation plan appropriate for an introductory class.

We now arrive at a major piece of the software design puzzle, the validation of our software. As a reminder, validation is the process of confirming that the right software was developed. This is not to be confused with verification ("the software was developed correctly"). Verification as a process checks the output of the software against the software specification. Here we close the loop and check the performance of our software against the actual user needs in the user's environment. This is the top layer of the V-model, shown in Figure 1.3.

Even though validation comes close to the end of the project, one should not begin the costly process of implementing and testing the software until there is a concrete validation plan in place. Otherwise many hours and much treasure could be wasted in developing something that has no hope of returning on the investment. In particular, if human or animal studies will be required, one needs to establish partnerships with the appropriate (often academic) hospitals and other institutions who can perform such experiments and to (at least examine whether it would be possible to) obtain approvals for such studies from institutional review boards.

The design of the validation study will depend on the claims we are making for our software. In many cases, the strategy for our validation of our system (either on its own or as part of a medical device) might be similar to that used in other scientific experiments. In others, it may simply be an extended set of testing at user sites.

Note: The validation step is where our software leaves the lab/office for the first time and migrates to a testing facility. At least part of the validation should be performed within the environment of the user using the actual software package that will eventually be released.

15.1 Regulatory Background

We restrict our discussion in this section to software as a medical device (SaMD), that is, standalone software that is not part of a device. The validation of medical devices raises more complex questions as these also have a hardware component. See Vogel [261] for more information on this topic.

This section begins with a reminder of some of the key regulatory terms (see also Chapter 2). Next, we will describe the regulatory guidance for SaMD (Section 15.1.1) in general. Finally, we describe emerging guidance for the validation of SaMD containing AI/ML modules (Section 15.1.2).

When beginning to think about validation, one always needs to remember the three critical issues that concern regulators (e.g. the FDA): (1) efficacy, (2) safety, and (3) security. In the case of software, part of safety and security is demonstrated "by design" and can be at least partially demonstrated through the system design documents. Our emphasis in this chapter will be primarily on efficacy, which is loosely translated as "better," or "just as good" (but cheaper). In general, the FDA defines efficacy in the case of a new treatment as showing that the treatment arm is "better" than the placebo arm or the standard of care, and "just as good" when compared to approved treatments. One needs to be mindful of these definitions when designing clinical trials. Most software validation, however, does not need this level of complication and often involves showing that the software works as intended in human study population or volunteers.

Unfortunately, these terms (better, just as good) are initially subjective/qualitative. To prove efficacy, what we really need are objective, quantitative evaluation criteria. Converting from qualitative to quantitative measures is a key task in this type of work.

Safety and Security For software the primary considerations here are usability – discussed in Section 13.5 – and cybersecurity – discussed in Section 3.5.

Independence of Review The other key concept from the regulators is the need for "independence of review." Figure 2.2 [92] illustrates when independent review (which is always ideal) is critical. As this shows, when the software risk categorization for a particular piece of software reaches the level at which software errors can result in significant harm, it becomes imperative that at least some of the clinical evaluation is performed independently by a separate team (or even by a separate company). This increases confidence in the evaluation of the software.

Independence of review is even more relevant to the validation of AI/ML techniques. Here, in addition, one should employ "independence of data." This means that the data used for testing an AI/ML module should be completely different from that used for training and tuning the original model, and even come from different sources [189].

Table 15.1 Clinical evaluation for SaMD. This summarizes the three components of clinical evaluation: (1) valid clinical association, (2) analytical validation, and (3) actual clinical validation. Reproduced from the FDA's SaMD Clinical Evaluation document [92].

Valid clinical association	Analytical validation	Clinical validation
Is there a valid clinical association between your SaMD's output and your SaMD's targeted clinical condition?	Does your SaMD correctly process input data to generate accurate, reliable, and precise output data?	Does use of your SaMD's accurate, reliable, and precise output data achieve your intended purpose in your target population in the context of clinical care?

15.1.1 Components of SaMD Clinical Evaluation

In Section 5.1 of the IMDRF/FDA clinical evaluation document the process is described as: "The assessment and analysis of clinical data pertaining to a medical device to verify the clinical safety, performance and effectiveness of the device when used as intended by the manufacturer" [92]. This process is divided into three components that we examine next – see also the summary table presented in Table 15.1.

1. Valid Clinical Association This is summarized in the document as: "Is there a valid clinical association between your SaMD output and your SaMD's targeted clinical condition?" [92].

Fundamentally, we must demonstrate here that the output of our software (e.g. measurement of tumor volume from MRI) is a valid measurement that is significant (useful) for the underlying clinical application (e.g. deciding on whether radiation treatment is needed). This may be accomplished by a variety of strategies. At one end, we may simply be able to rely on existing scientific literature and professional guidelines to show that this relationship (e.g. tumor volume as an appropriate input to a clinical decision process) is valid. This may be particularly the case where our software simply automates/improves the computation of some parameter that is already in use.

On the other hand, if our output is a new type of measurement we may need to create such evidence by performing either new experiments (e.g. clinical trials) or (if we have it) use existing data and perform new analyses to demonstrate that our proposed measurement has a valid clinical association with the target disease. In this second example, we may have a secondary feature of a tumor (e.g. shape or other image characteristics) that can be computed from existing images and can be used to retrospectively analyze tumor type (which has a direct impact on patient outcome) to see if it is useful for diagnostic and/or prognostic purposes. To clarify: A measure is diagnostic if it permits us to identify the disease/problem. A measure is prognostic if

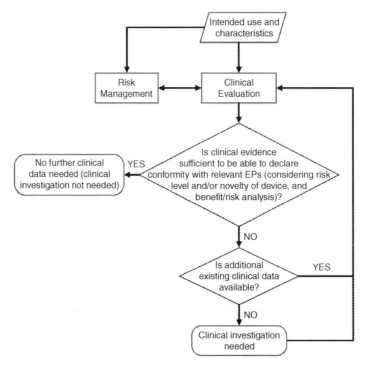

EPs = Essential Principles of safety and performance of medical devices;
* - Conformance to performance standards may be sufficient to demonstrate compliance to relevant Essential Principles.

Figure 15.1 When is clinical investigation needed? This figure shows a flowchart that can be used to decide whether a clinical investigation on a device must be performed. Reproduced from the IMDRF [130].

it can be used to predict the outcome of a treatment strategy. The first concerns the present, whereas the second has to do with the future.

On a related note: A recent, relatively brief, guidance document from the IMDRF [130] contains a useful flowchart (Figure 15.1) that explains when a new clinical investigation is needed.

2. Analytical/Technical Validation This is summarized as: "Does your SaMD correctly process input data to generate accurate, reliable, and precise output data?" [92]. Further, the document poses two questions:

1. Does the software reliably produce the accurate/correct outputs, given specified inputs?
2. Does the software meet its specification and user needs?

Combined, these represent the verification and (the technical) validation of the software. Was the software correctly implemented (verification) and was the correct software implemented (validation)? As the IMDRF states, "Analytical validation is necessary for any SaMD" [92].

3. Clinical Validation The careful reader will have noticed that we have already done validation in the last section. What more is needed? The last piece is to demonstrate that our software is actually clinically useful. What our users specified as their needs may in fact not be relevant. To again quote from the IMDRF document: "Does use of your SaMD's accurate, reliable, and precise output data achieve your intended purpose in your target population in the context of clinical care?" [92].

Our goal, then, is to demonstrate that the output of our software has a "clinically meaningful" output for our target use. Clinically meaningful is defined to mean that the output of our software has:

positive impact of a SaMD on the health of an individual or population, to be specified as meaningful, measurable, patient-relevant clinical outcome(s), including outcome(s) related to the function of the SaMD (e.g. , diagnosis, treatment, prediction of risk, prediction of treatment response), or a positive impact on individual or public health. [92]

There are two principal ways to accomplish this task. The first is to use or reference existing data from previous studies that demonstrates the clinical validity of our measurement. The second is to generate new clinical data via a study targeted at this problem.

A second related task has to do with what the document terms "Clinical usability/User interface considerations." Our software has to enable clinically meaningful output when operated by our target users. This will require some form of *human factors validation testing* defined as [89]:

Testing conducted at the end of the device development process to assess user interactions with a device user interface to identify use errors that would or could result in serious harm to the patient or user. Human factors validation testing is also used to assess the effectiveness of risk management measures.

To amplify this statement, this form of testing is a risk analysis measure (see Section 5.2) that helps identify hazards that may arise from the use of our software. In addition to safety improvements, this type of analysis may also be used to improve the performance of the software and to create a more pleasing setup for the user. We discuss this in more detail in Section 13.5.

To summarize: We must demonstrate that our output is a valid measurement for our target application (clinical association), our software produces the output correctly (analytical/technical validation) and that the output can be used to positively impact clinical outcomes (clinical validation).

We note that the FDA has domain-specific guidance for certain applications. For an example, see guidance from the FDA [84] that discusses issues related to the evaluation of computer-assisted detection algorithms applied to radiology images.

15.1.2 Validation of ML/AI-Aided Software

The introduction of AI/ML techniques presents some additional challenges to the validation of software.[1] An excellent introduction to the topic from a regulatory perspective can be found in a guideline document from the NMPA [189].[2] This is a brief document and should be read in its entirety.

The overall process of creating, evaluating, and deploying software that uses AI/ML modules is shown in Figure 9.3 in Section 9.3. In that section we discussed the processes of creating training data and selecting and training a model. These are steps A–D of the flowchart. Here, we continue the story and describe the clinical evaluation components – steps E and F.

Our discussion here is guided by material in Section 2.4.2 of the NMPA document [189]. This begins with a basic introduction to software validation that is similar to that presented in Section 15.1.1. What separates the validation of "AI-aided" software from regular medical software is the topic of training/evaluation data. The next two sections of this document discuss clinical trials (using data acquired specifically for this purpose) and retrospective studies (using existing data that was acquired/generated for a different study or as part of standard clinical routine). The document next discusses two validation settings: clinical trials and retrospective studies.

Clinical Trials We will discuss prospective clinical trials in the next section (Section 15.2) in more detail. What separates this type of evaluation from retrospective studies is that new data is being acquired as part of the validation of the AI module which contains our trained model. Our emphasis here is on the AI/ML aspects of the clinical trial.

The first important point made in the document is the recommendation that one should select a third-party central laboratory to evaluate clinical trial results. This addresses the same issue of independence of review that is discussed in the FDA/IMDRF clinical evaluation document [92] – see also our discussion in Section 2.2.3, and in particular Figure 2.2.

In addition, the advice is that the testing organization(s) should be different from that which was the source of the training data.[3] This is an attempt to evaluate how well the results generalize (see our discussion in Section 8.4) to different sites and equipment. Finally, they recommend that the number of institutions be as large as possible to further assess generalization.

Clinical evaluation for AI/ML software may involve more than just the performance evaluation of the AI/ML alone. It may also include an assessment of the clinician's performance using the tool (in the case of computer-aided diagnosis systems, for example).[4]

Naturally all other considerations for clinical trials, such as adherence to ethical constraints and selection of diverse populations, apply as well, as discussed later in Section 15.2.3.

Retrospective Studies To both facilitate medical device innovation and to avoid the cost of clinical trials, the NMPA [189] states that one could, for lower-risk software,[5] perform a retrospective study using existing data.[6]

The document states that these retrospective studies should be part of the initial design and that they should be appropriately controlled to avoid issues of bias, whether in data selection or in the evaluation criteria. The same recommendations with respect to the selection of data as in clinical trials apply here as well. The testing data should ideally come from institutions that did not supply training data. Further, again, a third-party organization should be charged with the coordination of the testing. This is key to ensuring the "accuracy, safety and effectiveness" [189] of the software.

15.2 Human Subject Studies and Clinical Trials

15.2.1 What is a Clinical Trial?

In many of the validation tasks described in the previous section, we may need to conduct a clinical trial.[7] The FDA has a whole set of guidance documents on conducting clinical trials that are beyond the scope of this introductory book. Our goal is to introduce the reader to some basic concepts. A good overview can be found in the book *Clinical Trials* [203] by Pawlik and Sosa.

The US National Institutes of Health define a clinical trial as:

A research study in which one or more human subjects are prospectively assigned to one or more interventions (which may include placebo or other control) to evaluate the effects of those interventions on health-related biomedical or behavioral outcomes.

The key word here is "prospectively." A prospective clinical trial is one in which the study design, outcome specification, and analysis plan (and any other critical parameters) are fully defined prior to any experiments and data collection. This avoids the temptation to adjust the conditions once the experimental process is underway, which may introduce biases in the process. A second bias-avoidance technique is randomization – see Section 8.6. The subjects (if we are testing multiple groups or *arms*, e.g. placebo and treatment arms) must be randomly assigned to each group to avoid biases in subject assignment.

15.2.2 Testing Medical Devices in Patients or Patient Data

If our software requires FDA approval or clearance (see Section 2.3), then it will fall under the broad category of medical devices. Our software may be part of a device or the whole device (e.g. a SaMD). Hence, the clinical trials process for this will follow the same broad FDA guidelines for devices.[8]

As described in Chapter 2 (see in particular Section 2.3), the FDA has different classes for medical devices with different requirements approvals. The 510(k) pathway [86] allows for new devices to be cleared if they are shown to be "substantially equivalent" to previously cleared "predicate" devices, and therefore may not need new human clinical trial data.[9] However, novel devices with no predicate often fall into the more laborious Premarket Approval (PMA) pathway [81].

If our device or software requires clinical data (so as to, for example, demonstrate performance improvements[10]), we are faced with a classic chicken-and-egg situation:

We need approval prior to clinical use and we need clinical use prior to approval! The solution to this problem involves some form of review, whether at the institutional level (all major medical schools have an institutional review board (IRB) that approves human studies[11]), or even from the FDA itself under the Investigational Device Exemption (IDE [79]) process. Obtaining an IDE is required for US human studies for a device that (1) has not been approved for the proposed task and (2) poses a significant risk to patients. There are two types of IDEs:

1. **Feasibility study:** This is often a small study whose goal is to establish the safety of the device and the potential for effectiveness. This should follow prior studies in a pre-clinical setting (e.g. animal studies and phantom studies). Feasibility studies are often "single arm" and do not have a placebo or comparator group.
2. **Pivotal study:** This is a larger study and serves as the primary clinical support for the full FDA application. This is designed to demonstrate a "reasonable assurance of safety and effectiveness." The endpoints and the number of subjects involved are driven by prior statistical analysis.

The FDA review of these is more involved. Pivotal studies are typically blinded and placebo (or comparator) controlled, which means patients are randomly put into groups to either receive study treatment or some comparator such as placebo or the medical standard of care. Additionally, both the investigators and patients should be "blinded" or unaware of which treatment they receive. Placebos are readily available as pills, but are more difficult in the case of medical devices.[12]

15.2.3   Issues in Human Subjects/Clinical Trial Design

FINER and PICOT FINER and PICOT are two acronyms that summarize much of the best practices in designing a research study. A good introduction can be found in the short paper by Farrugia et al. [66]. A fuller introduction to the topic can be found in the book by Hulley et al. [115]. As they state: "FINER denotes the five essential characteristics of a good research question: It should be **f**easible, **i**nteresting, **n**ovel, **e**thical, and **r**elevant" [115]. FINER is a useful tool for creating the big-picture question. The second acronym "PICOT" addresses the more detailed aspects of the design. PICOT [66] stands for **p**opulation, the patient population that we are interested in, **i**ntervention, the intervention to be investigated,[13] **c**omparison group, the main alternative to which we are comparing,[14] **o**utcome of interest, what we will measure, and **t**ime, the appropriate follow-up time after which we assess the outcome.[15] These acronyms are useful as summary guides in designing a study.

Ethical Considerations We begin this discussion by quoting from the famous Declaration of Helsinki, titled *Ethical Principles for Medical Research Involving Human Subjects* [270]. In particular, Article 8 of the declaration states:[16]

While the primary purpose of medical research is to generate new knowledge, this goal can never take precedence over the rights and interests of individual research subjects.

The IMDRF *Clinical Investigation* [130] document also has a short discussion (Section 7) where beginning from the quote above provides some additional guidance. Essentially, the

desire to protect human subjects from unnecessary or inappropriate experimentation must be balanced with the need to protect public health through the use of clinical investigations where they are indicated. In all cases, however, care must be taken to ensure that the necessary data are obtained through a scientific and ethical investigational process that does not expose subjects to undue risks or discomfort. [130]

A practical ethical consideration involves the process of subject recruitment. We must ensure that the subjects have provided "informed consent" for their participation in the trial. This process involves clearly explaining the potential risks and benefits (if any) to the patients and ensuring that they clearly understand what they are entering into. Informed consent forms (ICFs) are required to be signed by patients after detailed explanation is provided by study investigators.

Ethical considerations also enter into the determination of the study population. This is to both protect groups from harm and to ensure that the study populations are diverse enough to ensure applicability to both sexes, multiple ethnic groups, different age groups, etc. as appropriate – see Section 8.3.3 for a mathematical rationale for this. This is particularly important when validating AI/ML modules.

15.3 Designing a Validation Strategy

We focus here on the case in which one has to create new data/perform a new study to establish the validity of our software. The same principles apply to reanalyzing existing measurements.

15.3.1 Some Key Questions

Designing a validation strategy often begins with asking some of the following questions:

- Is there a current solution? If yes, does our new method do appreciably better (or cheaper, etc.)? Does it permit a less skilled user to perform the task than previous solutions? Can we demonstrate at least equivalent performance?[17]
- How does one measure "better" performance? What does it mean to be better in a statistical sense?
- Is this a new solution? Is the performance good enough? How does one measure "good enough"? What is the current standard of care?

While the questions are generic, the answers to these questions are domain-specific. The first time one tries to design a validation plan, there is often a sense of hopelessness/despair; since while the questions are obvious, they seem never to apply to the problem at hand. Often the details of how to measure things present an insurmountable obstacle. In general, however, having the right questions and patience (and experience) will often yield a reasonable (though never perfect) set of answers.

Some of the work needed for the validation plan properly belongs to the overall "needs-finding" phase of the project. Some of these questions should have been asked at that time:

- Ask the user what it would take for them to believe that your software works.
- Ask the user what it would take for them to buy your software.
- Convert qualitative and subjective statements into objective and quantitative criteria. Justify the quantitative criteria. (e.g. "achieving a 25 percent improvement in accuracy will enable treating a class of patients that cannot be currently treated").

Based on the answers to these questions we then design a set of experiments to create the measurements (potentially including a direct comparison with existing methods) to prove that our software works/is better.

Note: If this information is not available in the system requirement specification, then the validation engineer needs to "stop" the process and ensure that the system requirement document is revised to include it. Then and only then can we proceed with the design of a validation plan.

15.3.2 Some Additional Constraints in Software Validation

Unlike verification, in which we will often use standard software engineering techniques such as regression testing, software validation uses as its input the "finished" software package. At this stage in the process, we are not examining the source code or interacting with the code directly, but rather testing it from the perspective of the user.

In an ideal setup, the people (validation team) performing the validation ought not to have been involved in the implementation of the actual software. A real software validation test involves checking that the software meets the needs of the user, so we need to test the software in an environment that matches (as closely as possible) the user's actual environment.

15.3.3 Process

Example 1: The goal of our software is to enable a user to perform a standard task accurately and efficiently. The first steps in designing the validation plan are to:

- define the task precisely so that it can become a repeatable procedure;
- define accurately – for example, identical to the existing method, correct subject to manual checking, at an accuracy of 99 percent subject to some standard; and
- define efficiently – for example, in less than 2 minutes

The next step is to define the user in more detail (this should be in the system requirements). An example might be "The user is a nurse with some computational expertise using a standard PC."

The next step is to recruit a set of users that is representative of the population of users (the discussion in Section 8.3.3 is relevant here). We will ask the users to perform the task in a simulated yet realistic environment and measure their performance in terms of accuracy and efficiency. A friendly statistician would then be asked to analyze the data[18] to establish that our software meets or exceeds user expectations/needs in terms of accuracy and efficiency.

Example 2: The goal of our software is to enable a user to perform a certain task better than they could using the standard/existing method. The solution to this will involve some form of randomized study. We will have to divide our users (using randomization – see Section 8.6) into two groups. Group 1 performs the task using the standard method and Group 2 performs the task using our method. We then perform statistical hypothesis testing to demonstrate that our method produces statistically significant improvements over the standard method.

15.3.4   An Example Validation Requirements Template

For each combination of user and use case of the software, we need to design a procedure to validate the software. You can think of each section as answering a question beginning with the words "What? Why? How? How will you know? What if it does not work?"

Goal or hypothesis (what?): Our description starts with the statement of a goal or hypothesis: To demonstrate that the new technique significantly outperforms an existing technique by improving accuracy by 30 percent. Our hypothesis/goal needs to be specific and quantitative.

Rationale (why?): This section justifies the hypothesis/goal and basically answers the question "Why is this a good thing?" For example, the rationale here could be that improving accuracy by 30 percent allows us to perform certain procedures that we could not do otherwise. This recapitulates (in shorter form) material from the system requirements.

Methods (how will this be done?): What experiments will you perform? This should detail the experimental procedure in recipe format and end in a set of quantitative measurements. Use figures and perhaps even a flowchart to illustrate the procedures clearly.

Data analysis (how will we know that we have succeeded?): How you will analyze your measurements to prove your hypothesis, thus demonstrating successful validation of the software. In those cases that involve showing improved performance, the most common statistical analysis technique is some form of hypothesis testing (which may involve t-tests or an ANOVA). For those cases where the goal is to show equivalent performance to an external gold standard (e.g. algorithm vs. expert human), the various possible analysis techniques include computing correlations, using Cohen's κ [35] and Bland–Altman plots [20].

Alternative strategy (what if this does not work?): What could go wrong and what is a potential plan B?

A common mistake in designing this plan is to treat the data analysis section as an afterthought. This is a mistake. The data analysis section is what will enable you to prove that your experiment succeeded. Writing it may require consultation with a bio-statistician to ensure that an adequate number of measurements is made and to design the appropriate statistical procedure. This is particularly important in those cases in which one needs to demonstrate improved performance compared to existing techniques. It is not sufficient to simply demonstrate improvement; it must also be shown that this improvement cannot be explained by chance (or any factors unrelated to the new procedure) alone. The key aim here is to obtain statistically significant improvements. The concept of statistical significance is discussed in Section 8.5.

15.4   Conclusions

The validation of medical software is one of the last steps in the process prior to applying for regulatory clearance. While this comes at the end, one should not begin the costly process of implementing and testing the software until there is a concrete validation plan in place. The validation of medical software has three components: (1) ensuring that the output of the software is relevant to the targeted clinical condition; (2) ensuring that the software generates accurate, reliable, and precise outputs; and (3) that these outputs are useful in the context of clinical care. The work required for each of these steps will vary depending on both the novelty and the risk categorization of the software. An additional level of complexity comes with the use of AI/ML modules. The validation of the performance of these brings additional requirements to ensure separation of training and testing data.

15.5   Summary

- Software validation is the process of ensuring that our software meets the needs of our user as captured in the system requirements document.
- Validation is performed using the finished software package in an environment that is similar to the (or the actual) environment that the user will ultimately use the software in.
- The regulatory guidance breaks the process down into three steps: (1) valid clinical association, (2) analytical validation, and (3) clinical validation.
- For the purposes of clinical validation, our goal can be one (or a combination of) demonstrating that it has "good" performance, "better performance than some alternative" or "equivalent performance to some alternative" at lower cost (or requiring lower operator skill level/training).
- As part of this process, we may need to contact either a prospective clinical trial or a retrospective clinical study. Both are governed by regulations to ensure the safety and privacy of the study subjects.

- Prospective clinical trials that involve the recruitment of new subjects have significant ethics rules to ensure that the patients involved are properly informed about the risks involved in the study and that the clinical research does not take precedence over the rights and interests of the research subjects.

RECOMMENDED READING AND RESOURCES

For a regulatory perspective, see the FDA's classic document:

FDA. General principles of software validation; final guidance for industry and FDA staff, January 11, 2002.

Our description follows mostly the more modern guidance that can be found in:

FDA, Center for Devices and Radiological Health. Software as medical device (SAMD): clinical evaluation. Guidance for industry and Food and Drug Administration staff, December 8, 2017. This is a re-issue of IMDRF/SaMD WG/N41.

For the validation of AI modules, see the NMPA (China) document titled:

NMPA. *Technical guideline on AI-aided software*, June 2019. We used (and verified) an unofficial translation made available through the website of the consulting company ChinaMed Device (https://chinameddevice.com).

The following study makes the case that medical AI software that was approved by the FDA is often insufficiently validated due to the lack of multi-site testing and prospective studies :

E. Wu, K. Wu, R. Daneshjou, et al. How medical AI devices are evaluated: limitations and recommendations from an analysis of FDA approvals. *Nat Med*, 27(4):582–584, 2021.

A good overview of clinical trials can be found in the book:

T.M. Pawlik and J.A. Sosa, editors. *Clinical Trials*, 2nd ed. Springer, 2020.

For a description of clinical research techniques, see:

S.B. Hulley, S.R. Cummings, W.S. Browner, D.G. Grady, and T.B. Newman. *Designing Clinical Research*. Wolters Kluwer, 2013.

P. Farrugia, B.A. Petrisor, F. Farrokhyar, and M. Bhandari. Practical tips for surgical research: research questions, hypotheses and objectives. *Can J Surg*, 53(4):278281, 2010.

SAMPLE ASSIGNMENT

Based on your system specification document (the assignment for Chapter 12), create an outline validation plan for your project. The document should be approximately 1–2 single-spaced pages long. Please submit a draft of this document a week prior to the deadline for peer review.

NOTES

1. An additional challenge involves the validation of software that adapts and improves over time using new data. The FDA discusses this in a recent draft guidance document [82], which we briefly discuss in Section 2.6.

2. We reference here an unofficial translation that was obtained from the website of the consulting company ChinaMed Device [30].

3. For example, if the training data came primarily from Hospital A, the testing site should ideally not include Hospital A.

4. This may be in addition to standalone testing on independent data to benchmark performance and provide a better understanding of a potential biases or subgroup concerns.

5. Each regulatory zone uses a slightly different categorization. One example based on the recommendations of the IMDRF [126] is shown in Table 2.1.

6. This is a recommendation by this particular regulatory authority and may not apply to other regulatory zones.

7. This section describes the process in the United States as specified by the FDA, though much of the material is applicable internationally.

8. The information in this section is derived in large part from the FDA's website and in particular a presentation by Owen Faris. The FDA is constantly updating its policies and requirements. It is worth checking for updates when creating an approval plan. For example, at the time of this writing the FDA "Enforcement Policy" for digital health devices treating many psychiatric disorders permits clearance without human clinical trials during the COVID-19 pandemic. In other words, the FDA occasionally does not enforce its own policies during periods of public health emergency.

9. Sometimes, though, new human data is needed. The FDA uses multi-reader, multi-case clinical studies to assess many computer-aided detection (CADe) and computer-aided Diagnosis (CADx) devices even though they are class II. These studies could be retrospective (using old data) or prospective (using data acquired explicitly for the study).

10. As the FDA notes [79]: "clinical studies are most often conducted to support a PMA. Only a small percentage of 510(k)s require clinical data to support the application."

11. Many studies happen at major medical centers and are approved by such IRBs. There are also industry-based IRBs that approve studies that companies perform on their own. The FDA has overall oversight of IRBs.

12. The process for drug studies is similar. These also proceed from Phase I (safety testing in a small group) to Phase II (larger group divided into control and experimental groups, to determine safety and effectiveness) to Phase III (often large and multi-site trials), which are the last step prior to requesting approval from the FDA to introduce the drug to the market. Phase IV trials can be performed after approval to evaluate long-term safety.

13. This might be the use of new algorithms to better estimate tumor size from MRI.

14. Such as the current state-of-the-art or older software.

15. We are interested in the actual impact of the new technique on patients' health outcomes, not just software/algorithmic performance metrics such as accuracy.

16. This declaration was written in reaction to the crimes committed in Nazi Germany in the name of "medical research." See also *The Nuremberg Code* [192].

17. This is important for predicate-based regulatory clearance such as the FDA's 510(k) process [86].

18. The statistician may not be so friendly unless he or she was asked to help design the experiment, including, critically, the number of subjects (users) prior to the measurements!

16 Deployment, Maintenance, and Decommissioning

INTRODUCTION

In this short chapter, we discuss the final three steps in the software life cycle: (1) deployment (Section 16.1), (2) maintenance (Section 16.2), and (3) decommissioning (Section 16.3). These are complex topics in their own right. Our goal here is to simply and briefly highlight some information about each of these processes so as to make the reader aware of what the major issues are for the sake of completeness.

16.1 Deployment

We have now reached the promised land. Our software is complete and we are ready to move to the exciting phase of the deployment of our software to our users. Our discussion follows here the discussion in Section 8.5 of the IMDRF QMS [127] and Section 5.8 of IEC 62304 [119]. For medical software in particular, this can be a complex process and requires the completion of a number of steps, ranging from ensuring that the verification process is completed to checking that the documentation is finalized. This last step includes documenting any anomalies in the software, the version number, how the software was created, that the software release is archived, and that we can ensure the software can be reliably delivered [119].

We first need to define the term *anomaly*, which the standard defines as "any condition that deviates from the expected based on requirements specifications, design documents, standards, etc. or from someone's perceptions or experiences. ANOMA-LIES may be found during, but not limited to, the review, test, analysis, compilation, or use of SOFTWARE PRODUCTS or applicable documentation" [119].

Essentially, what this is saying is that prior to deployment we need to make sure that everything is in order. In particular, we need to know if there is something that is "anomalous," or irregular, about our software that makes it deviate from our requirements and design.

While the above applies generally to all software, when our software is (or is part of) a medical device, there are additional complications. Deployment may include several steps:

Delivery and installation: This is the process of actually getting the software to the user. In some medical software, delivery includes a preconfigured workstation with the software pre-installed. In other cases, we may need to ensure the software is correctly installed on a workstation at the site (whether new or existing).

Setup and configuration: Our software will need to be appropriately configured to function in the user's environment. This may include interfacing with hospital IT (for example) to ensure that network/database access is appropriately set up. The settings for the software need to be documented and clarified.

Repeatability: We need to have a process that permits repeatable software deployment. This should include delivery, installation, setup, configuration, intended operation, and maintenance [127].

User documentation and training materials: These should identify limitations of the software, and other information that should be considered during deployment. The documentation should include information to make sure that the installation is correctly done and procedures for the installer/user to verify that this was in fact done correctly. Please note that the documentation and associated training materials for the software is considered part of the software itself [122], and should be created with the same level of care as the product itself.

Risk management: The customer needs to be made aware of hazards identified during the development process and any residual risks that remain. Failure to inform the customer of potential hazards can lead to catastrophic results, as illustrated in the vignette presented in Chapter 21 describing fatal accidents involving the Boeing 737 MAX aircraft.

In conclusion: Our software is potentially a medical device. It is not simply something that one downloads, installs, and hopes that everything goes well! We need to ensure that our user has all the information necessary (and potentially hands-on support) to install and configure the software and be aware of its limitations and any risks associated with its use.

16.2   Maintenance

The maintenance of medical software is a complex process in its own right. In fact, IEC 62304 has a flowchart very similar to that shown in Figure 10.2 for the maintenance process – this is figure 2 in the standard, which on first inspection looks like a duplicate of figure 1 (which we adapt in Figure 10.2). In the maintenance process, in addition to the ever-present risk management step, we have the following steps [119]:

- **Establish software maintenance plan**: This replaces software development planning in the original "creation" process.
- **Problem and modification analysis**: This replaces the requirements analysis in the original process.

- **Modification implementation**: This consists of steps that are completely analogous to what we have seen before, that is, software architectural design, software implementation, and testing.

In summary, software maintenance is, in many ways, a repetition of the original design process. The key change is that our input is not user needs (and by extension requirements), but problem and modification analysis. This consists of a number of steps as well.

The first step is to establish a process for maintenance. This will include, naturally, procedures for receiving, documenting, evaluating, resolving, and tracking any issues reported by users (external or internal) [119] and processes for acting on such reports. We will need to determine whether an issue is a problem, evaluate this for risk, and if necessary activate procedures to make changes to the software.

A particularly critical step is to evaluate any problem reports that relate to safety. In addition, any requests for changes need to be evaluated carefully for the effects the change may have on the software and any external systems with which our software interfaces. Users often fail to appreciate that what appear to be small changes may require massive changes to the underlying software.[1] Any change requests need to be evaluated and approved, and any changes must be communicated to both users and regulators. A similar process must be followed to disclose new cybersecurity threats and mitigation strategies. Once the changes are implemented, the re-release of the software follows a very similar path to that described in Section 16.1.

Note: Maintenance and software updates can be a major source of problems. Part of the problem is that in many cases, the original developers of the software are no longer with the company, and new software developers need to make changes to the existing code. This is always a challenge and should be appropriately accounted for in the risk management plan.

16.3 Decommissioning

At some point, our software reaches end-of-life status and we will no longer support it. This could be because it has become obsolete or because it has been replaced by a substantially different newer version that does not qualify as an upgrade. Sometimes this is also because manufacturers decide to exit a particular business for a variety of reasons.

At the end-of-life stage, the software needs to be decommissioned as it will no longer be supported/maintained. The goal of a good decommissioning process is to "stop the maintenance, support and distribution of the software in a controlled and managed fashion" [127]. This process needs to be slow and deliberate. The users need to be given plenty of warning that the software is coming to the end-of-life phase and be, ideally, provided with support to both extract any valuable data from the system and transition to a new system. We must also ensure that data privacy is safeguarded and any data still on the system is properly erased prior to the hardware being recycled or repurposed. This is all good business practice as well in terms of retaining potential customers for new products.

16.4 Conclusions

Deployment, maintenance, and decommissioning are the last steps in the software life cycle. Our goal in this chapter was to briefly highlight the importance of these steps. Their presence in the life cycle underscores the seriousness of the medical software process. Software must be appropriately deployed (installation, integration, user training). We are required to have mechanisms to receive user feedback and take the necessary corrective action as needed. Finally, the process of retiring the software must ensure that our clients are supported in transitioning from our tools and that appropriate measures are taken to ensure data privacy.

16.5 Summary

- Medical software is often deployed to the user site (e.g. a hospital) by the manufacturer. This includes both installation, integration with existing software, and appropriate user training. The documentation of the software (and the training materials) are an integral part of the whole.
- The process of software maintenance directly parallels the process of designing and implementing the software itself.
- A particularly important aspect of software maintenance is the process of documenting, evaluating, resolving, and tracking any issues reported by users. Failure to do this may have serious consequences, as in the case of the Therac-25 incidents (see Chapter 17).
- A major challenge in software maintenance is the fact that the developers maintaining the software may not have been involved in the original design and implementation. Care must be taken, therefore, to ensure that the software is implemented in such a way that makes this process possible.
- When the software reaches the end of its lifetime, the manufacturer should support users in transitioning to a new solution and ensuring that data privacy is maintained.

RECOMMENDED READING AND RESOURCES

The two best sources for this chapter are:

IEC 62304 Medical device software – software life cycle processes, May 2006.

IMDRF, SaMD Working Group. Software as a medical device (SaMD): application of quality management system, October 2, 2015.

For more on the Therac-25 incidents and the failure of the manufacturer to react to customer complaints, see:

N.G. Leveson. *Safeware: System Safety and Computers.* Addison-Wesley, 1995.

NOTE

1. The opposite is also true. What appears as a major change to a user may in fact be a very simple change for the software team. Communication is key in this process.

Part IV

Case Studies

17   Therac-25: Software that Killed

Vignette written with Ellie Gabriel

INTRODUCTION

This vignette describes the Therac-25 radiation therapy machine, whose software bugs and lack of hardware safety interlocks led to six serious accidents between 1982 and 1987. Three lives were claimed due to overconfidence in software and loose design regulations. Ultimately, these events were a catalyst for the FDA to begin investigating and regulating medical software.

17.1   Radiation Therapy

17.1.1   Introduction

External beam radiation therapy is an effective treatment option in cancer. It works by either killing the cancer cells to treat the disease or shrinking a tumor to ease the symptoms.[1] A machine aims beams of X-rays or electrons at specific parts of a patient's body. Radiation is something all individuals experience when they have X-rays taken of their teeth at the dentist or naturally when they stand close to a microwave.[2] X-rays, made of photons (light particles), are generated by using a high voltage to accelerate electrons from a hot cathode into an anode; the subsequent rapid deceleration produces X-rays.[3] Although electron beams do not involve photons, this is a type of ionizing radiation that disrupts the chemical bonds of DNA, thereby breaking the DNA beyond use. In effect, the energy of the radiation destroys the genetic material in the cancerous cells to the point where they can no longer grow, divide, and threaten the patient [185].

17.1.2   Complications from Radiation Therapy

Improvements in radiation therapy have allowed physicians use high-dose radiation to preferentially target cancerous cells while sparing healthy tissue to a larger extent than before. However, there is still damage to normal cells that results in common side-effects such as hair loss at the treatment site, fatigue, fertility problems, and nausea. The precise radiation dose is chosen to maximize the effect on cancer cells and minimize the harm to healthy cells. Radiation treatments are usually performed multiple times a week, with time allowed for healthy cells to recover. Electron therapy is preferable to

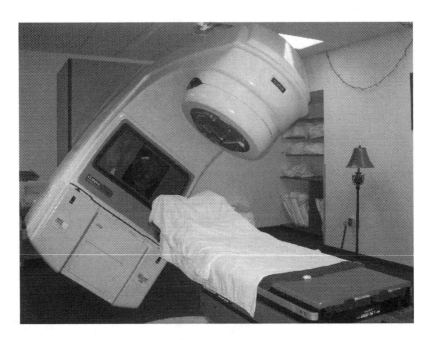

A modern radiotherapy machine similar to the Therac-25. The picture shows a device designed to treat cancer much like the Therac-25 machine described in this vignette. Image courtesy of Catalina Màrquez, obtained from Wikipedia.

X-ray therapy for superficial cancer because electrons release the bulk of their energy at the tumor site and stop there, while X-rays release most of their radiation immediately upon entering the body and continue on past the tumor, leading to more frequent and severe side-effects of the therapy [268].

Radiation is created using a linear accelerator (LINAC) – see Figure 17.1. When this is not self-shielding, patients must wear lead shields to block healthy tissue areas from severe damage. Overexposure to radiation can result in radiation poisoning, whereby too many healthy cells are destroyed, weakening bodily functions. The patient may develop chronic illnesses such as more cancer. Unintended extreme radiation in a single dose can cause death [166].

17.2   The Therac Accidents

17.2.1   The History and Design of the Therac-25

The best description of the Therac-25 accidents can be found in Appendix A of the book *Safeware: System Safety and Computers* [154] by N. Leveson.[4] Atomic Energy of Canada Limited (AECL) designed a computer-controlled radiation therapy machine called the Therac-25 in 1982. It was nominally simple and economical in design, and much of its functionality depended on software controls.

In the case of the Therac-25, the LINAC produced electrons that could then be applied directly in electron mode (after spreading using scan magnets to reduce the dose to safe levels). In photon mode, the electrons (using 100 times greater energy) would be directed at an X-ray mode target, which would produce photons that subsequently would be filtered using a flattener to reduce doses to acceptable levels. The power level (electron vs. photon mode) and the filter (scan magnets vs. X-ray target and flattener) were controlled by independent processes. In particular, a turntable was used to place either the scan magnets or the X-ray target and flattener in the path of the beam. In addition, the device had a light-field mode in which, instead of radiation, it produced visible light to help a technologist align the beam position with the patient.[5]

Unlike previous Therac models, which had hardware such as safety interlocks and radiation intensity monitors to ensure the radiation mode and the turntable were in agreement, the Therac-25 relied exclusively on software control. To complicate matters, the engineers of the Therac-25 reused software from these older models in a ritual called "cargo cult programming," in which they blindly copied over old code. It was later discovered that the hardware safety features of Therac-6 and -20 masked their software bugs without indicating that any faults even existed. Disaster occurred on six separate occasions, where a high-energy electron beam generated in X-ray mode was delivered directly to the patient (i.e. high energy without the X-ray target and flattener).

17.2.2 Description: One of the Accidents

Here is a description of one of six recorded accidents (East Texas Cancer Center, April 1986 [154]). This was a case of skin cancer that was to be treated using electron radiation. The technician incorrectly chose X-ray mode before quickly (within 8 seconds) changing to electron mode by clicking the keyboard. Due to a software error, the electron beam was still set for X-ray mode, without the proper shields in place.[6] The equipment did not properly synchronize with the technician interface (see Figure 17.2), so that race conditions occurred.[7] In this particular situation, the technician essentially instructed the computer to simultaneously prepare for both X-ray mode and electron mode, which resulted in an error. The message "Malfunction 54" appeared on the screen, but technicians fell into the habit of pressing "p" to proceed and bypass this error, which was not in the Therac-25 user manual. This caused the patient to be directly hit with a very high dose of radiation (a high-energy beam meant for X-rays but without the X-ray target and flattener). The patient died from this radiation overdose three weeks later.

17.2.3 Results of the Investigation

AECL initially denied physicists' claims that the injuries and deaths could possibly be caused by their machine, and complaints were brushed aside.[8] After more investigations, in 1987 AECL informed physicists that the FDA had approved their corrective action plan (CAP) to fix all machines that year. This required over 23 software

```
PATIENT NAME:  TEST                                                    A
TREATMENT MODE: FIX              BEAM TYPE: X ENERGY (KeV):             25

                                ACTUAL         PRESCRIBED
            UNIT RATE/MINUTE       0                200
            MONITOR UNITS        50  50             200
            TIME (MIN)           0.27              1.00

GANTRY ROTATION (DEG)             0.0               0              VERIFIED
COLLIMATOR ROTATION (DEG)       359.2             359             VERIFIED
COLLIMATOR X (CM)                14.2             14.3            VERIFIED
COLLIMATOR Y (CM)                27.2             27.3            VERIFIED
WEDGE NUMBER                                       1              VERIFIED
ACCESSORY NUMBER                  0                0              VERIFIED

DATE :84-0CT-26        SYSTEM: BEAM READY      OP.MODE: TREAT       AUTO
TIME:  12:55. 8        TREAT: TREAT PAUSE               X-RAY       173777
OPR ID:T25V02-R03      REASON: OPERATOR        COMMAND:
```

Figure 17.2 Text-based user interface used in the Therac-25. The radiation therapy technicians of the Therac-25 machine described in this vignette used a text-based terminal interface to enter setup information and prepare the machine. In order to understand the accidents, one must note that the terminal lacks an organized user interface but instead is entirely text-based. Image from Leveson and Turner 1993 [156]. © IEEE. Used with permission.

changes and at least six hardware safety features, including a dose-per-pulse monitor that physicists had demanded. Old-fashioned hardware saved the future Therac-25 machines, though the device's reputation is still marred by the past [156].

17.3   Regulatory Reforms and Future Considerations

The Therac-25 disaster was a watershed in the formation of the FDA and of the agency's regulatory responsibilities. Essentially, the FDA began to consider medical software as a potential source of serious problems in safety-critical systems after it saw the capabilities of the Therac-25. In 1990, the Safe Medical Devices Act (SMDA) was enacted, requiring users to report medical device-related accidents to the FDA and the device manufacturer. Previously, only the manufacturers were required to make reports, which led to a lack of information exchange [72].

Additionally, concerns arose about the ability to validate software. It is virtually impossible to check every possible keyboard pattern and software behavior, especially for a major program like the Therac-25, which involved over 100,000 lines of code [219]. It is necessary but not sufficient to demonstrate confidence in a software's validity, efficacy, and overall safety; hence, to offset the infinite flexibility of software (and inherent dangers), mechanical safety interlocks need to be put in place. The particular coding error is not as important as the generally unsafe design practices, over-reliance on software, and overconfidence in the Therac-25 that a software engineer should always keep in mind.

In a follow-up paper, 30 years after the accidents, Leveson [155] reviewed the lessons from these accidents. In her conclusion, she stated:

My original account of the Therac-25 losses said that accidents are seldom simple. They usually involve a complex web of interacting events with multiple contributing technical, human, organizational, and regulatory factors. We aren't learning enough today from the events nor focusing enough on preventing them. It's time for computer science practitioners to be better educated about engineering for safety.

NOTES

1. The National Cancer Institute has a nice introduction to this topic [185, 186].
2. Note that this is low-dose radiation.
3. See the article "X-ray production" on Radiopaedia [211].
4. This builds on the original paper by Leveson and Turner [156].
5. One may consider this visible-light based alignment as an early form of augmented reality technology.
6. This scenario probably never occurred during testing prior to release as the software testers typically carefully follow prescribed routines in a calm and stress-free setting. In the real world, however, we see users under time pressure make mistakes and try to correct them, which is what ultimately caused the error.
7. A condition in multi-threaded software in which two parts of the program access the same piece of memory, leading to unpredictable results. Race conditions are hard to debug as the issue only appears in very specific situations.
8. Appropriate response to feedback from users and especially reports that pertain to patient safety is a critical aspect of software maintenance. The speed and methodology of this event reporting has evolved in recent years and the advent of the internet has led to the public often finding out quickly and rejecting a form of therapy altogether.

18 Mars Climate Orbiter: Lost without a Trace

Vignette written with Ellie Gabriel

INTRODUCTION

This vignette describes the accidental destruction of a multi-million-dollar outer space satellite, primarily due to inconsistency of mathematical units used in different components of the system. Poor fault analysis, decision-making, integration testing, and auditing led the Mars Climate Orbiter to burn up in Mars' atmosphere, never to be contacted again.

18.1 Outer Space Exploration

18.1.1 Introduction

Spacecraft are machines specifically designed for traveling through outer space. Spacecraft have many purposes, such as transportation of people or items, communication, observation, and collection of items from other planets. Satellites are launched into space attached to launch vehicles like rockets, which set the spacecraft on the right initial path; once a spacecraft is in space, ground control can make small corrections to its course. Scientists can determine where a spacecraft is by sending a radio signal with a time code on it up to the machine. The satellite bounces the signal back to Earth at a particular angle at the constant speed of light; based on how long it takes for the signal to return, scientists can determine how far away the object is – a similar technique is used by satellite GPS. The inertial property of spacecraft ensures that they will continue moving through space unless forced not to. As such, small deviations from course at the beginning of the flight will accumulate to much larger ones as a craft travels a long distance, such as to another planet.

18.1.2 The Start of the Small Spacecraft Initiative

With the rising costs of space missions in the early 1990s, the National Aeronautics and Space Administration (NASA) began seeking out more economical scientific missions.

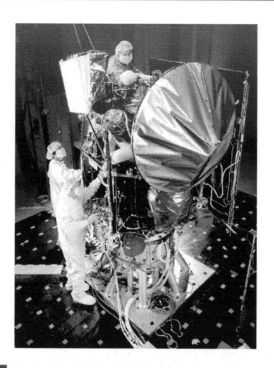

Figure 18.1 The Mars Climate Orbiter. This is the craft described in this vignette, pictured here undergoing acoustic tests that simulate launch conditions. Image courtesy of NASA.

In 1994, NASA established the Panel on Small Spacecraft Technology for this purpose, and the Panel sent its first spacecraft to Mars in 1996 to measure geology. The next mission was to be the Mars Climate Observer, shown in Figure 18.1, which would study climate and weather on Mars [184].

18.2   The Mars Orbiter Mission

18.2.1   Launch and Initial Software Problems

The Mars Climate Orbiter was launched in December 1998 from Cape Canaveral Air Force Station in Florida. For the first four months of navigation to Mars, calculations of angular desaturation maneuver (AMD) forces were performed manually rather than through NASA's software; this was due to file format errors and random unfixed bugs in the newly implemented small forces model subroutine that estimated minor forces for extra accuracy. NASA therefore relied on information received from Lockheed Martin via email (!) to set the AMD forces such that the spacecraft was rotating properly. These AMD forces worked to counteract external pressure and consequential turning. The manual calculations were in fact accurate [163].

18.2.2 Trajectory Correction Maneuvers

On the way to Mars, flight controllers navigated the spacecraft by using the Doppler shift in the radio link to measure motion and position. They were able to calculate the flight path ahead of time and could even "target" particular changes in the spacecraft's motion and command the rocket engines to adjust in what is called a trajectory correction maneuver (TCM). Without a clear and accurate understanding of the forces acting on the spacecraft, it would be difficult to determine the actual orbit and to accurately target trajectory changes.

18.2.3 Final Destination

After nine months, the time came for the Mars Climate Orbiter to stow away its solar panels as it prepared to descend into the upper atmosphere of the red planet. It was to reorient itself to line up its engine with Mars, ignite its main engine burn, pass behind Mars, losing contact with ground control for 21 minutes, and enter an elliptical orbit around Mars; a series of aerobraking maneuvers over the next two months would put the spacecraft into a circular orbit. To everyone's dismay, the Mars Climate Orbiter never re-established contact after those nail-biting 21 minutes, and the satellite was likely entirely burned up in the atmosphere that it was not supposed to enter at all.

The minimum altitude that the satellite was thought to be able to sustain was 80 kilometers from Mars' surface. Post-failure calculations demonstrated that the orbiter (when it was lost) was likely around 57 kilometers from the surface (170 kilometers off-course), where it would have been at extreme risk of violently skipping on the atmosphere and burning up without a trace. This is described in the NASA report *Mars Climate Orbiter Mishap Investigation Board Phase I Report* [178].

18.3 Fault Investigation

18.3.1 Asymmetry in Design and in Information

Two fatal flaws are cited: First, on the spacecraft's journey to Mars, propulsion maneuvers were in place to prevent the craft's wheels from gaining too much angular momentum. These momentum wheels, approximately 10 centimeters in diameter, are spun up and down by electric motors up to 3,000 revolutions per minute – think of a gyroscope. These can steady the spacecraft against unsettling torques with fine control as compared with gas jet thrusters. Unfortunately, the asymmetric arrangement of solar panels on the craft increased the solar sail effect that it experienced (Figure 18.2). In this effect, photons from the Sun exert a rotating force (photon pressure) on the aircraft, which would have been a relatively small concern had the spacecraft not been traveling a distance of 196 million kilometers. Ultimately, AMDs were necessary to counteract

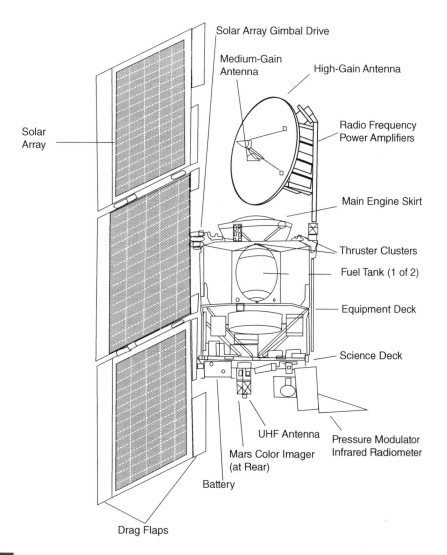

Solar Array Gimbal Drive

Medium-Gain Antenna

High-Gain Antenna

Solar Array

Radio Frequency Power Amplifiers

Main Engine Skirt

Thruster Clusters

Fuel Tank (1 of 2)

Equipment Deck

Science Deck

UHF Antenna

Pressure Modulator Infrared Radiometer

Mars Color Imager (at Rear)

Battery

Drag Flaps

Figure 18.2 The design of the Mars Climate Orbiter. This blueprint [184] shows the asymmetric design of the Mars Climate Orbiter. Note the three solar panels on one side, which increased the solar sail rotating effect.

this force, which made the wheels spin too fast and require slowing down by the force of small gas jet firings. These AMD events happened 10–14 times more often than expected due to photon pressure, inducing a large build-up of momentum. Further, with the asymmetrical design of the spacecraft and the off-center center of mass, each thrust of the small gas jets increased imbalance. In this cross-coupling of forces between rotational axes and pure translation, the spacecraft is moved away from its original course while it is being rotated as desired (see Figure 18.3).

Second, the Lockheed Martin ground control software that calculated the AMD force returned a value with imperial units (pound-seconds). NASA's software, however, expected a metric value (Newton-seconds) as outlined in their coding standards.

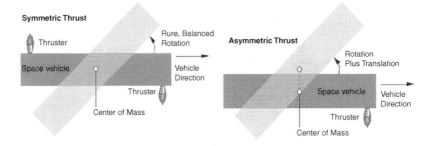

Figure 18.3 Asymmetry in the Mars Climate Orbiter. To understand how the Mars Climate Orbiter went so far off-course, one must understand how the asymmetric design led to translation along with rotation each time the small gas jet thrusters were fired in an attempt to minimize the rotation caused by the solar sail effect. Original figure from Oberg [193] © IEEE. Used with permission.

This discrepancy, off by a factor of 4.45, was compounded over millions of miles. Each time, the ground control computers were told that the spacecraft had received a force more than four times as great as it actually had. The difference in calculated and measured position led to a discrepancy between desired and actual trajectory. Ironically, these measurements were critical for correcting the first fatal flaw. NASA needed to know from JPL how to update the system with the calculated course change throughout the mission. To do so, scientists at JPL calculated the translational force induced by each rotational jet during the AMD maneuvers and multiplied it by the observed time duration. Again, while the error in units led to large measurement discrepancies over time, the fact that the units are different by a mere factor of 4.45, meaning that they are of the same magnitude, suggests that the unit differences were not particularly alarming to the system engineers – the imperial and metric units are close enough in magnitude – at least before being compounded.

18.3.2 Failure to Address

A handful of sources also suggests that overconfidence and carelessness within the NASA team led them to miss opportunities to rescue the mission before disaster ultimately struck. For instance, individuals concerned about the trajectory calculations were dismissed because they did not fill out formal complaint reports. In the face of doubt, a team should make a serious effort to prove that all is right rather than force the questioners to prove that something is wrong. Sources of unease were never resolved. However, experts like Thomas Gavin, deputy director for space and earth science at NASA's Jet Propulsion Laboratory, and Carl Pilcher, science director for solar system exploration at NASA headquarters in DC, agree that while it played a large role, human error is not the only major lesson here. A unit conversion should not be allowed to bring down a multi-million-dollar spacecraft. The control system failed to detect and correct errors during the mission. This is a lesson in risk management and error control [193].

In NASA's official *Mishap Investigation Phase I Report* from 1999, it is noted that the "wide-spread perception that 'Orbiting Mars is routine'" [178] led to complacency and inadequate consideration of risk mitigation. They also note a significant untimeliness and uncertainty surrounding the critical flight decision to plan a TCM-5, which could have potentially flown the spacecraft out of the danger zone.

18.4 Lessons Learned

Following the Mars Climate Orbiter incident, NASA recommended better training of navigation staff, a priori (what could go wrong) analysis of fault trees, contingency plans with firm decision criteria, software audits for specification compliance on all shared data, delegation of responsibilities to team members deciding when to execute maneuvers, and an audit of all software for the use of imperial units. NASA, rather than placing responsibility on Lockheed alone, attributes the lost mission to a failure of checks and balances in their own software. In addition, the recommendation was that the organization must be cognizant of heeding early warnings, not brush them aside. NASA implemented these suggestions in 2005 when it built and launched the Mars Reconnaissance Orbiter, which has successfully completed most of the intended goals of the Mars Climate Orbiter [178].

The Mars Climate Orbiter would be the first controlled flight into terrain (CFIT) accident in space, whereby a perfectly well-designed machine is flown into the ground – or turbulent atmosphere – due to human error in ground control [193]. Safety is a quality that must be established and re-established under new conditions. In healthcare, it is imperative that safety systems are in place for the earliest warning systems possible, given there is usually a patient on whom the technology is utilized.

19 HealthCare.gov: The Failed Launch of a Critical Website

Vignette written with Ellie Gabriel

INTRODUCTION

This vignette discusses the failed launch of the online marketplace for purchasing individual healthcare insurance that was created as part of the Affordable Care Act (ACA) in fall 2013. A key component of this effort was the website www.HealthCare.gov, through which individuals would be able to buy health insurance. The launch of this website failed, resulting in significant problems in the implementation of the ACA. This was a major process failure: software best practices were not followed, resulting in major cost overruns.

19.1 Background

The provision of healthcare in the United States relies primarily on the purchase of health insurance plans from private insurers. These plans are, for the most part, group plans that are made available as part of employee benefits. One of the major efforts to reform this system, the ACA, was a bill enacted in 2010. The primary component of the ACA was the creation of HealthCare.gov, a website intended to act as a virtual marketplace for purchasing individual healthcare insurance policies by those individuals who did not receive them as part of their benefits packages. These individuals are typically either low-paid or self-employed workers. HealthCare.gov was mission-critical in the sense that the goals of the ACA rested on its success, with much of the success hinging on the ability of the general population "signing up online" for a health plan. This "mission" of building up the website, unfortunately, was executed in a flawed manner and two years of work crashed down in a mere two hours [99].

Figure 19.1 shows a picture from an address given by President Obama in October 2013 during which he alluded to the issues with launching the HealthCare.gov website.

19.1.1 Pre-Launch Obstacles: Unheeded Warnings

The website designers received numerous critical reviews from outside consultants who noted over a dozen risks of launching the site, pointing out inadequate planning

Figure 19.1 President Barack Obama speaks about the ACA. This speech was given at Faneuil Hall in Boston, Massachusetts on October 30, 2013, one month after the launch of Healthcare.gov. Image: Boston Globe / Getty Images.

for online traffic flow as well as general inconsistencies with usual web design standards. The Health and Human Services (HHS) Secretary herself, who oversaw the process, hired another consultant to preview the website and suggest improvements; interestingly, the advice of this consultant was never shared with the website designers to be implemented in any way. HHS's Centers for Medicare and Medicaid Services (CMS) dismissed technical warnings about issues that customers would go on to experience after launch. Additionally, the team never considered delaying the launch of the website in light of these warnings. Again, the problem was overconfidence that the system would work and the site would not crash, despite some people warning the traffic might be too great.

The website itself was not completely built; therefore, it could not be tested thoroughly. The parts simply were not all connected – a major red flag. The crude organization of the website developed from poor decision-making and team management very early on in the project.

In terms of funding, the ACA provided the HHS with $1 billion. This money was to be used to implement the ACA, but over half of this grant was channeled toward the Internal Revenue Service (IRS), among other agencies to carry out the bill. Tension within the White House due to a changeover from a Democratic to a Republican majority in the House of Representatives led to micromanagement of minor issues such as official phrasing, increasing delays in progress. By fall 2013, the HHS team had become desensitized to cautionary reports concerning its progress [99]. It was overly optimistic about a major undertaking, leading to dangerous tunnel vision.

19.1.2 Risks and a Lack of Risk Management

While the CMS had organized Medicare and Medicaid programs in the past,[1] the HealthCare.gov service would prove a complex challenge – one that no one was quite prepared to coordinate in a short time period. Additionally, uncertainty about funding and constantly changing policy led to disorganized project direction. A significant amount of time that should have been spent building, testing, and refining the website was instead spent resolving issues within the government body. This bureaucratic culture clashed on multiple occasions with the innovative startup mentality common to web developers. The time that the contractors actually spent building the website was limited to a few months. CMS granted funding to the website developers in 2011, but it did not specify details about how the website would need to perform until 2013. This condensed time frame only increased the likelihood of design flaws and ultimate failure. Contractors did not have adequate time to complete the project, but they were still paid upwards of $2 billion, as opposed to the estimated initial cost of approximately $200 million. CMS issued cost-reimbursable contracts with the web developers, meaning that the developers would be paid whether they completed the project or not. These high-risk contracts led to financial strain on HHS and little contract control. Lastly, there was a distinct lack of senior leadership. The White House chief technology officer was not included in the HealthCare.gov project, making it even more challenging for the web developers to navigate the political issues involved [153].

Inadequate risk management held the team back, delaying the project further. Rather than moving forward, CMS would allocate resources to make up losses. Without contingency plans in place to identify potential risk management strategies ahead of time, CMS was working blindly. All things considered, CMS officials simply refused to adjust project plans when setbacks occurred – they had no risk plan to follow, amplifying each setback along the way. Risk mitigation strategies are especially important in large projects spanning multiple agencies and departments. With no clear, effective leader to assess progress, recognize problems, decide priorities, and clearly be able to halt and solve the errors prior to launch, the project was doomed [153].

19.2 The Launch of the Website

19.2.1 Launch-Day Failure

On October 1, 2013, HealthCare.gov was officially launched. After two hours it crashed. The root of the issue was immediately apparent: 250,000 users attempted to view and use the website at the same time. This was five times more traffic than was expected on the website. The site was not designed to handle this incredibly high demand, leading it to crash. Besides the traffic issue, there were additional flaws in design and functionality, primarily because the website was not complete at the time that it was launched. Rather than waiting to perfect and thoroughly stress-test their site, the US government chose to rush their biggest healthcare project of the decade. One of the immediately apparent issues that website users faced was incomplete drop-down

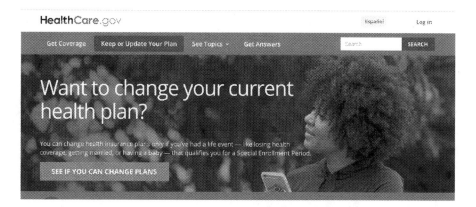

Figure 19.2 The HealthCare.gov website in 2021. This website is the federal online marketplace that allows individuals to purchase healthcare insurance policies in the United States.

menus; insurance companies themselves complained that they received incorrect or incomplete user data through the website. Ultimately, a mere six people were able to select plans on HealthCare.gov on launch day [3]. This number slowly grew to 26,794 over the course of October, but was still only 10 percent of what the federal government had planned for [153].

19.2.2 Remediation

After the initial crash, the government scrambled to recruit talent from Silicon Valley and other top technology companies in a "tech surge." This effort, along with a newfound interest in teamwork between government officials and IT workers at a central command center in Virginia, helped repair the website [99]. By the end of 2013, approximately four in five people were able to successfully use the site; however, the disastrous initial attempt significantly obstructed the implementation of the ACA [153].

The first health plan enrollment season ended on April 1, 2014. Immediately after this day, HHS leaders met to strategize for the second sign-up season that would begin in the fall. They considered the website's numerous defects and ended up prioritizing approximately half of them as "actionables." Problems remained, however, for years after. For example, during the third enrollment season, people faced challenges using the HealthCare.gov website, reporting incomplete functions [153]. A recent screenshot of HealthCare.gov is shown in Figure 19.2.

19.3 The Post-Failure Investigation

19.3.1 Summary

The HHS conducted a two-year comprehensive review of the HealthCare.gov failure in an effort to discern what went wrong with the website design and what was done to solve these problems. The 84-page study is primarily based on interviews with employees of "HHS, CMS and companies that worked on the project, as well as on several thousand emails, memos, government contracts and other internal documents" [47].

The HHS's Office of Inspector General notes that the failure can be mostly attributed to a series of disregarded warnings. Clearly, government "officials failed to recognize the magnitude of the project, were disorganized, were disadvantaged by shifting ACA policies, had too little money, used poor contracting practices, and finally, ignored problems until it was too late" [47].

The report firmly concludes for the public that the problems occurred not because of the design and IT work, but rather because of mismanagement by the health officials attempting this large feat. The project overseers needed to improve their organization more than their technological skills. Stronger leadership and clearer responsibilities within a project will help speed up decision-making in the future, as well as ensure that problems are recognized and addressed in an organized manner [47].[2]

19.3.2   Software-Specific Issues

We highlight here several software-specific issues as discussed in the aforementioned case study [47]. These illustrate that, while one can have a "proper" process in theory, failure to adhere to the process can lead to failure overall. The development of complex software is very much a "team exercise" and all members of the team must play by the rules.

"The selected method of software development made it easier for policymakers to seek frequent and late changes": The contractor, CGI Federal, proposed to use an Agile development method – see Section 9.2.2 for a description. In Agile methods the requirements are not completely specified at the outset of the process (such as happens in the Waterfall and other sequential models – see Section 9.2.1), but rather an iterative process is followed where some requirements are identified, built, and tested for a specified period of time (a "sprint" [225] or "increment" – see Section 9.2.2), and then the next set of functionalities is worked on. An advantage of Agile development is that each increment results in a functioning product and, as such, permits constant evaluation of the product. The disadvantage, in this case, was that:

The business owners (in this case, CMS staff) responsible for a particular process can assess whether the software meets the project's needs and adjust business requirements accordingly. This ability to adjust, however, enabled policymakers to frequently change business requirements and technical specifications on an ongoing basis. Changes made through the Agile development process must still be properly considered, documented, and communicated. Managers at CGI Federal reported in interviews that the frequency of CMS's requests for change resulted in too much change too late in the process, contributing to delays. For example, CGI Federal managers reported that CMS did not define business requirements at the beginning of each sprint, and often made changes throughout the sprint, which inhibited the Agile method and resulted in incomplete development.

There was an agreed-upon life cycle (Agile), but the business owners failed to adhere to this. This is a classic out-of-control scenario that violates all notions of design controls. There was a process, but it was not followed.

"Compressed timeframe for technical build": As stated in the report, the "final months of development and implementation for HealthCare.gov were chaotic for CMS staff and contractors. The nine months from January to September 2013 provided, from the outset, very little time to accomplish the tasks remaining." This included "tasks critical to success, such as testing website functionality and security, and ensuring adequate capacity for users" [47]. Part of the reason, as mentioned above, was that CMS made changes to the "specifications well into 2013, delaying development to a point where it was not feasible to complete and test the website as initially planned."

"CMS senior leadership failed to fully grasp the poor status of the website build, and to alter its course": This is organizational failure. As per any quality management system (see Chapter 4), the managers responsible need to have the necessary competence to manage a project.

"CMS and contractors recognized they would not finish system functionality testing before the launch, but prioritized delivering the product on time over testing and resolving problems": This is enough to make any software engineer shudder. Testing is not an optional part of the development process.

"CMS prepared to launch HealthCare.gov on October 1, 2013, as planned, optimistic in spite of problems and never seriously considering delay": This is a recurring theme in many past failures. The external demands (whether corporate or political) create pressures to take actions that lead to disasters. In this case, the implementation of the ACA was so politically significant that it led to the managers of the project overruling technical concerns, leading in turn to the expected disaster.

There is more in the report that is well worth reading, including the story of how the project recovered by following improved procedures.

NOTES

1. Medicare and Medicaid are federal programs that provide health insurance for the elderly and for the poor, respectively – see Section 3.2.1.
2. It is also evident that people in this case underestimated the cost of such a project, as often happens in software overhauls and website design.

20 The 2020 Iowa Caucus App: An Unreliable App that Caused National Embarrassment

Vignette written with Ellie Gabriel

INTRODUCTION

On Monday, February 3, 2020, the results of the Iowa Democratic Party Caucus, the first caucus in the US election cycle, were delayed as a result of bugs in the software used to report the results. This software (the IowaReporterApp) was not properly tested prior to the elections, and the whole reporting system failed during the event. This led to multi-day delays in reporting the results of the caucus, throwing the entire primary election process into confusion and disarray.

20.1   The US Presidential Election Primary Process

20.1.1   Caucuses and Primaries

In the United States, presidential elections are preceded by a primary process during which the main parties select their candidates. Each state holds either a full primary election or a caucus. Caucuses, while only held in less than a dozen states as of 2020, have been around since the beginning of American politics. The first caucus every election year, which is the opening round of the presidential election, traditionally has been held in Iowa. In the Iowa Democratic caucuses more specifically, individual precincts hold meetings of registered Democrats – see Figure 20.1 for an example. Voters discuss their support for different candidates, perhaps touching on points that were brought up in the most recent candidate debates, and then each person casts a vote. The precinct leader tallies up the votes in his or her precinct to determine the winner. This winner's name is reported to the Iowa Democratic Party (IDP).

Equipped with this information from each precinct, the IDP can determine how many of its delegates to the Democratic National Convention will be representing each candidate. For instance, if there were an overwhelming majority of precincts in favor of Candidate A over Candidate B, then the majority of the delegates from Iowa would be sent to the Democratic national convention in support of Candidate A, with the remainder for Candidate B. Many states have replaced the caucus procedure with a simple primary election in which individuals secretly cast their votes for their

Figure 20.1   The 2008 Iowa Democratic Party Caucus. One can see signs for different candidates to which the voters align (sup-port); the specific candidates are supposed to walk in order to be counted. Image courtesy of user Citizensharp – Own work, public domain, https://commons.wikimedia.org/w/index.php?curid=3590252.

preferred presidential candidate, rather than having to attend a group meeting and vote by precinct. Unlike primary elections, which are run by the state governments, the caucuses are run by the parties themselves.[1]

20.1.2   The Political Technology Field

Interestingly, the very niche voting technology sector generates approximately $300 million in annual revenue. This is a fragmented market with a few big vendors and a handful of smaller ones – the lack of standardization is due to the fact that the requirements governing election equipment and technology vary by state. "Election equipment" need not be limited to actual polling of the public. It may include sending messages to potential supporters to promote a campaign, recruiting volunteers via their smartphones, collecting donations, and gauging public opinion on political issues.[2]

20.2   The 2020 Iowa Caucus

20.2.1   The Company

The company that developed the IowaReporterApp was originally called *Groundgame*; it was renamed *Groundbase*, *Shadow*, and finally *Bluelink*; this lack of secure identity should have been a red flag. The company was going bankrupt in 2019, having trouble finding its footing in the political technology field. Luckily for the company, Acronym, a non-profit digital outfit with affiliations in political action, invested a great deal in Shadow.[3] One year later, Shadow found itself in the middle of a firestorm.[4] Shadow's services had been used in the past by different Democratic candidates' political teams to send messages to supporters and solicit donations; note, however, that some of these teams halted use of the company amid concerns about security and data reliability [239].

20.2.2   The Plan

The Iowa Democratic Party (IDP) decided to revise the caucus procedure for 2020. It planned to record votes from the Iowa caucuses via a smartphone application in the hopes of making the process easier and faster in their 1,700 precincts. Precinct leaders would record the votes from their regions in the app and proceed to send the collective results to the IDP. The application would take care of the tedious calculations that typically require mathematical tables to allocate delegates. If, for some reason, the precinct leaders had trouble sending their results to the IDP, a back-up phone line was put in place for phone-in reports [263].

20.2.3   The Caucus

The caucus was scheduled for Monday, February 3, 2020. During the weekend prior, precinct leaders were sent a final update of the IowaReporterApp (Version 1.1 [146]). Simply put, this mission-critical application was still being updated 48 hours prior to the start of the 2020 presidential election year. Before voting even began, some precinct leaders stated that they had trouble downloading or logging into the app, but the IDP reported that these issues were mainly in areas of bad cell service or involved minor problems that would be resolved. These were some of the early warnings that the system would malfunction, as is often the case with initial minor issues.

On the Monday, the caucuses began as planned, but frustration and confusion swept through Iowa quickly after. Precinct leaders found the application difficult to download and access. Their authentication codes were being rejected, resulting in their phone screens shining vague error messages in their worried faces. The users were not trained either to use the application or to troubleshoot. Determined precinct leaders dialed in to the reporting hotline set up by the IDP. Because the application was failing across the entire state, phone lines were severely jammed. Precinct leaders who were successful in getting through were put on hold for upwards of two hours, while some calls were simply dropped for others [239]. This only added to the frustration during such a time-sensitive event. In one of the counties, only 2 of the 22 caucus leaders were able to successfully work the app.[5] It was a nightmare, and the whole country, which was eagerly waiting for results in front of their televisions, watched it unfold.

20.2.4 Partial Results Spread Doubt

While there was general difficulty downloading and accessing the app, the biggest concern was the coding error within the software itself that led to the system collating the application-reported information but only spitting out partial results. This was discovered late on Monday, and it would delay the final count by more than a day and raise questions. The IDP chairman made it clear that the coding error was in the back end of the software and that the raw voting data was coming in securely and accurately. He stated that no cybersecurity breach had occurred and that paper records would corroborate the accuracy of the data [239]. Still, in an extremely tense political atmosphere, delays in reporting results spread doubt about the integrity of the process.

20.3 The Software Engineering Aspects

20.3.1 Distribution Issues: TestFlight and TestFairy

The IowaReporterApp was not deployed through the traditional App Store(s)[6] that people are used to. It was installed through TestFlight [36], which is a way that app developers can release beta versions of applications directly to test users. This bypasses the typical App Store review processes (which check for hiding malware). Some businesses also use TestFlight to release applications to ensure they can run on more outdated, older operating systems without sluggish results. Nevertheless, whenever a user downloads an application through TestFlight or the Android equivalent, TestFairy, a warning message will appear on the phone, asking the user to confirm that they wish to download said potentially insecure application. An application intended to securely collect and report caucus results should not make smartphone users skeptical.

The decision of Shadow to release the IowaReporterApp through TestFlight and TestFairy is questionable in itself, and shows that the application likely did not undergo the rigorous testing that one would expect of a mission-critical application.[7] Shadow used these beta-testing platforms for general public release as a shortcut. This was neither the proper environment nor the right platform. TestFlight is not intended for releasing software at scale because it is less stable. In considering TestFairy, Shadow did not choose to purchase the premium developer plan; instead, it used the free plan, which deletes application data after 30 days and limits the number of users that can access the app to only 200. One should note that this "critical project" cost the IDP a mere $63,000. According to experts, the application should have cost at least double that for such a high-stakes project [241].

20.3.2   App Testing History and Issues

The app took approximately two months to develop. Questions have been raised about whether or not it was put through standard technology tests before being rolled out.[8] Security experts warned the IDP that the application had neither been stress tested enough nor put under a state-wide simulation. Even the cybersecurity chief of the Democratic National Committee urged Iowa not to use the application. The acting secretary of the federal Department of Homeland Security said that the department offered to test the app but that the request was declined [38]. A claim was made by the IDP that the app was tested by "independent cybersecurity consultants." Apparently, these "tests" that were done on the application occurred only a few weeks prior to the Iowa caucuses, and they were kept quiet [239].[9] The IDP did not want hackers to have time to figure out how to get into it. In addition, the app was never tested in the field, and there was not serious deployment process testing (people need to be able to download and log onto the application even in the most rural areas without a hitch).

20.4   Conclusions

One of the consequences of this fiasco was that many people began calling for Iowa to be demoted from its status as first caucus in the election cycle. After seeing what happened in Iowa, Nevada,which held its State Democratic caucuses later that month, decided not to use the reporter app it had paid Shadow $53,000 to develop. The political world learned the lesson that good software takes time and money to develop correctly and that shortcuts lead to disasters. A trilemma often applied to software describes what a potential user would like to have:

1. The software must have lots of features.

2. The software must be of high quality.

3. The software must be cheap to develop. This translates to fast development and reduced testing time.

As one can easily verify, only two of these three characteristics can be present. For example, a high-quality, feature-laden software package will not be cheap to develop and test.[10] In this case, what was sacrifice was quality, the one characteristic that one can never sacrifice in mission-critical applications. The trade-off should always be between features and cost.

To summarize, we will quote one of the experts interviewed by the *Washington Post* in the aftermath of the incident [239]:

Dan Wallach, who runs Rice University's computer security lab, said the lack of transparency erodes trust. "There was demonstrably not adequate training and testing," he said, pointing to apparent connectivity problems in rural parts of the state.

NOTES

1. A good explanation can be found in the article "How do caucuses work?" that appeared on the website How Stuff Works [113].
2. See the article 'Testing could have prevented Iowa caucus app failure, experts say' [162] by Angus Loten that appeared in the *Wall Street Journal* on February 4, 2020.
3. See the article 'The shoestring app developer behind the iowa caucus debacle' [98] by Glazer et al. in the *Wall Street Journal* on February 5, 2020 [98].
4. See the article "Fortune, Shadow Inc.: How a company with 120 Facebook likes ended up at the center of the Iowa caucus firestorm" by Morgan Enos, that appeared in *Fortune* on February 6, 2020 [58].
5. See the article "The app that broke the Iowa Caucuses was sent out through beta testing platforms" by Nick Statt that appeared on the website The Verge on February 4, 2020 [241].
6. The Apple App Store or the Google Play Store.
7. A proper test plan would have involved full field testing simulating actual election conditions, especially in areas of poor internet connectivity.
8. See the article by Corasaniti et al. in the *New York Times* titled 'App used to tabulate votes is said to have been inadequately tested' [38].
9. The expression in the security industry is that they were practicing "security through obscurity." This is usually looked down on as a tactic.
10. A similar trilemma applies to a healthcare system as a whole. The three desired characteristics are: (1) high quality, (2) low cost, and (3) fast service (i.e. low wait times). Only two of these can be present at a time at most. In reality, one can have a system that has only one or zero of these characteristics!

21 The Boeing 737 MAX Disasters: Using Software to Fix Hardware Problems

Vignette written with Ellie Gabriel

INTRODUCTION

In the case of the Boeing 737 MAX disaster, the manufacturer tried to address hardware problems with software fixes in order to avoid the costs of recertification of what would have been a new airplane (had the fixes been done in hardware). Overconfidence in the software and insufficient testing and pilot training led to two fatal crashes and 346 fatalities. Boeing has already paid a substantial fine to settle some of the legal issues arising from the case, though other legal procedures are still ongoing.

21.1   Background

21.1.1   The Boeing 737 MAX

Boeing is one of the largest aircraft manufacturers in the world. Many private airline companies purchase and use Boeing airplanes around the world. The workhorse plane for many airlines is the Boeing 737. In fact, some airlines such as Southwest Airlines [257] only operate (as of 2020) Boeing 737 aircrafts. This aircraft is a narrow-body jet with a seating capacity of 100–200 passengers, depending on the model. Around 2010, Boeing was rushing to develop the 737 MAX after hearing that its major rival company, Airbus, was unveiling the new A320neo plane, which would fly more efficiently than the then-current iteration of the 737, the, for our purposes, confusingly named 737 Next Generation. Boeing needed to develop the 737 MAX in just under five years to be ready to compete with Airbus. None of Boeing's planes up to that point had been made in such a short time frame. Many shortcuts would have to be taken [27].

The Boeing 737 MAX model has two flight-control computer systems that complement the pilot while in flight. They may provide full auto-pilot or simply some adjustments during manual flight to reduce the pilots' workload. These flight-control systems have greater power over major parts of the aircraft than the pilot, meaning that they can be difficult for human pilots to troubleshoot in cases of software system malfunction. Thoroughly tested hardware will always prove much safer than novel software that may have hidden bugs. A critical mistake that Boeing made was re-using the computer systems of the 737 Next Generation model. This model, while considered the "safest narrow-body airplane ever made" [27], had fewer system requirements than would be needed in the newest 737 MAX model.

The A320neo promised to fly farther with less fuel compared to Boeing's offerings. To match its rival, Boeing added new engines to the MAX that were larger and heavier than those on the previous model. Ultimately, the new engines changed the overall airflow around the plane. When flying high at steep angles, the newly disrupted airflow would cause the control wheel (think of the wheel of a car) to go slack, preventing the pilot from controlling the nose of the plane and its altitude. To fix this problem, Boeing could have added adaptive surfaces on the engines (allowing pilots to control altitude more easily), remodeled the wings, or used a stick pusher (a control that pushes on the control wheel at the right time to keep it from going slack). These hardware solutions would cost time and money – two things that Boeing was not prepared to give up with Airbus closing in on a critical segment of its business [27]. In particular, any changes to the actual physical airframe would have needed a thorough review by the Federal Aviation Administration (FAA).

21.1.2   Software to the Rescue: Maneuvering Characteristic Augmentation System

Boeing's solution was to use software instead. Changes in the software could be approved quickly and with little regulatory friction.[1] The software solution was called the Maneuvering Characteristic Augmentation System (MCAS) (Figure 21.1). When a 737 MAX plane entered a high incline, with the angle determined by a sensor outside the plane, the MCAS would activate in the background, rotating the plane to counteract the forces pushing the jet nose up [27]. Boeing told the FAA that the new MCAS was only a few lines of code and would barely need to be used except in extreme cases. Engineers on the project focused on other elements of the plane that they deemed more safety-critical. Boeing gave MCAS a "technical-hazard rating of 'major' implying that if it failed, it was not likely to cause destruction of the plane or passenger deaths" [248]. (The reader is directed to Chapter 5 for a more detailed description of risk management.)

However, the MCAS sometimes malfunctioned, misfiring over and over, sending the plane "into an infinite loop of nose-dives" unless the pilot was able to recognize the error, hit the emergency button, and begin damage control maneuvers within a mere four seconds. To add to the catastrophe, Boeing chose not to train their pilots on how to react to MCAS malfunctions, if it even told them MCAS existed at all – the word "MCAS" did not appear in the manuals [27].[2]

21.1.3   Other Design Issues: "Frankenstein"

One retired airline pilot, when looking at the Boeing 737 MAX design, described it as a "Frankenstein design." Rather than recognizing the potential for failure by design, Boeing used a flight-control system software band-aid to cover up the flaws. This band-aid

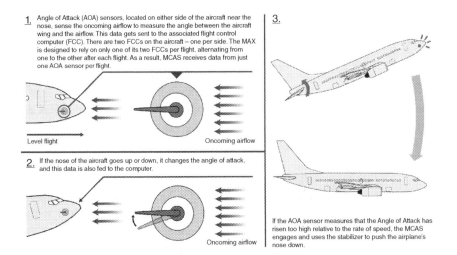

1. Angle of Attack (AOA) sensors, located on either side of the aircraft near the nose, sense the oncoming airflow to measure the angle between the aircraft wing and the airflow. This data gets sent to the associated flight control computer (FCC). There are two FCCs on the aircraft – one per side. The MAX is designed to rely on only one of its two FCCs per flight, alternating from one to the other after each flight. As a result, MCAS receives data from just one AOA sensor per flight.

Level flight   Oncoming airflow

2. If the nose of the aircraft goes up or down, it changes the angle of attack, and this data is also fed to the computer.

Oncoming airflow

3.

If the AOA sensor measures that the Angle of Attack has risen too high relative to the rate of speed, the MCAS engages and uses the stabilizer to push the airplane's nose down.

Figure 21.1 Boeing 737 MAX physical diagram. The new, larger engines on the MAX models causes increased pitch, making the plane tilt upwards when it should not. MCAS was developed to counteract this behavior; the angle-of-attack sensor on the front of the plane signaled the MCAS to activate and put pressure on the horizontal stabilizer to push the nose back down. Figure from a report by the Office of the Inspector General of the US Department of Transportation issued on June 29, 2020 [54]. Department of Transportation, Office of the Inspector General. Timeline of activities leading to the certification of the Boeing 737 MAX 8 aircraft and actions taken after the October 2018 Lion Air accident. Report AV2020037. June 29, 2020.

was MCAS [227]. Unfortunately, most pilots were not told how to use MCAS. By not instructing their pilots in the software that they would be working with or how to troubleshoot it, Boeing was setting its pilots up for disaster.

On the outside of the Boeing 737 MAX planes is an "angle of attack sensor." When the plane is pointing its nose too high, the MCAS system is activated to push the jet nose back down. In the 17 years leading up to the incident, there had only been 27 angle-of-attack alerts made during more than 240 million flight hours. Boeing treated this as an extremely rare event and as such did not notify pilots, hence most of them did not even know that the MCAS system was running on the 88 ton planes they were in control of [56].

The MCAS system itself was fallible in that it had only one sensor, one source of input. Without safety checks and balances, this was worrisome but never mediated [227]. Boeing felt that additional sensors would add unnecessary complexity to their grand new design [248]. Incorrect signals from the malfunctioning sensor told the computers that the plane's nose was too high and needed to be pushed down. In reality, this sent the plane into a nose-dive toward the ground at high speed [248].

21.1.4 Limited Pilot Training

In 2014, Boeing released a statement that pilots who were already flying 737 models would not be required to complete a simulation course for the new 737 MAX planes, despite the software changes [202]. Boeing elected to test the MCAS system only with veteran pilots,[3] who had more experience quickly troubleshooting malfunctions, to save time and ultimately money. Note that not all Boeing planes are flown by US-based airlines. For instance, Lion Air and Ethiopian Airlines both use Boeing planes but are based in Indonesia and Ethiopia, respectively. Other countries have different regulations for training their pilots.

The MCAS system on its own was not sufficient for successful flight. The proper training of pilots was necessary as well. There was too much faith that the software system would function flawlessly and that the pilots would know what to do in the unlikely event of failure [56]. Boeing engineers assumed that pilots would react correctly 100 percent of the time by using the emergency procedures that they knew for the problem of runaway stabilization, wherein the nose of the plane also dips too low [248]. Boeing "overestimated pilot skill while underestimating the probability for software errors" [56].[4]

During the development stages of the 737 MAX, Boeing carefully removed all references to MCAS in official manuals. Boeing also worked hard to keep FAA officials from criticizing their MCAS system or potentially finding the flaws, to the extent that Boeing referred to MCAS by other names. While the effort to cover up flaws is clearly on Boeing's end, the FAA was blamed for majorly insufficient oversight efforts and failure to identify major safety issues [202].

21.1.5   Increasing MCAS Power: Fuel to the Fire

After initially submitting documentation to the FAA about MCAS, Boeing elected to give more power to the system. Test pilots noticed that the control wheel did not stiffen as it needed to during low-speed maneuvers. For the plane to operate more smoothly, engineers would need the plane to experience increasing amounts of pressure as the pilot pulled back on the wheel. To make this adjustment, Boeing quadrupled the amount that the system could move the horizontal stabilizer to increments of 2.5 degrees rather than 0.6 degrees. This change allowed the automatic MCAS to force the plane's nose down much faster and more powerfully than before. After making this change, Boeing did not submit a new safety report to FAA, taking away the potential opportunity to find safety problems. Boeing continued to assume that pilots would infallibly know how to react to a nose-dive situation even though now the stakes would be much higher [248].

21.2   Disaster Strikes

Two Boeing 737 MAX airliners crashed on separate occasions, killing a total of 346 passengers. The first crash, an Ethiopian Airlines plane, occurred in March 2019. The second was in October of that year through Lion Air [56]. While in flight, the MCAS system malfunctioned, "repeatedly and aggressively" pushing down the jet nose, overriding pilot commands, and sending the plane head first into the ground. The pilots, not knowing they even had this software feature installed, failed to remedy the situation and counteract the system misfirings [202]. Before they began grounding their planes, Boeing concluded that the problem lay in the training of crew members. In maintaining that MCAS complied with safety rules and regulations, Boeing attempted to minimize the scrutiny over the MCAS system by deflecting the problem onto the airlines instead. The training records of the co-pilot in the Lion Air flight showed that he was unable to recite troubleshooting practices that he was expected to have memorized. In addition, Lion Air as a whole had a questionable safety record, unlike other airlines. The 322-page report by the Indonesian National Transportation Safety Committee notes that Lion Air continually flew Boeing 737 MAX planes even after many failures in airspeed and altitude had occurred [56]. After the Lion Air crash, the FAA continued to put little weight on the matter of an MCAS redesign, despite having agreed with Boeing that the MCAS needed one. Planes remained in flight [202].

Despite the public outcry caused by the two deadly crashes, Boeing continued to assert that their own standards of regulation were sufficient. Boeing sought to further persuade the FAA to decrease training requirements for the 737 MAX on the whole to make its technical evaluation of the planes congruent with the FAA expert views [202]. Eventually, however, all Boeing 737 MAX planes were grounded. Production was halted. Boeing spent approximately $5 billion to compensate passengers who had booked flights on the grounded planes and $50 billion more to compensate the families of the victims of the crashes [56]. The FAA proposed a $19.7 million penalty against Boeing for having installed unapproved sensors on hundreds of its aircraft, including the 737 MAX planes. This allegation demonstrated Boeing's failure to observe its own safety control standards.

21.3 Disaster Aftermath

21.3.1 Initial Fixes

Initially,[5] Boeing expected to roll out software fixes for the MCAS within a few months. Unfortunately, investigation of the flight control software led to even more safety concerns. Ultimately, much more analysis and testing was needed before Boeing 737 MAX models were assuredly safe. Flaws in the reliability of technical hardware as well as a failure to back up software changes increasingly slowed down Boeing's process. Additionally, the software used for flight-simulation training of pilots may have fallen out of date from current software expectations, compounding Boeing's inability to remedy the problems started by the Boeing 737 MAX models in a timely manner. One year following the Lion Air crash, Boeing ran into problems with a new software function intended to fix the MCAS. The new software prevented the flight-control computers from starting up and verifying that the plane was ready for flight. By not powering up correctly, the entire computer system ended up crashing. Prior to the identification of this flaw, most proposed software fixes were tested in ground-based simulators where no power-up problems arose [201].

Boeing recognized that it needed to both remedy its software issues as well as develop better pilot-training standards [202]. This was expected to be a slow-going process as long as Boeing continued to insist that it could fix the model's flaws with more software rather than opting to add another computer. For reference, the successful Airbus A320neo that prompted the hastily designed 737 MAX has seven computers compared with the MAX's two [27]. Boeing did add a second sensor to double the amount of input being fed to the on-board computers. It also planned to share more information about MCAS with the pilots and lessen the power of the system over the pilot. Additionally, it will no longer rely on textbook pilot reactions in its risk management plans [248]. Meanwhile, the government will use the investigative reports to provide momentum for legislative changes in the FAA to provide stricter safety protocol [202].

21.3.2 Regulatory Response and Implications

The FAA's reputation for conscientiousness had been damaged by the 737 MAX disaster from the start. Notably, the substantial influence of Boeing over the FAA is in part because many of the FAA regulating responsibilities were delegated to Boeing employees authorized to act on behalf of the government [202]. After having been pressured by Boeing to certify the MAX before it was ready, the FAA is now determined to not make the same mistake again. Boeing's planes have been placed under greater scrutiny than ever before, which has led to the uncovering of hardware issues that also do not meet FAA standards. Simply gaining FAA approval of the MCAS software will certainly not be the final step before the 737 MAX models can return to the air [27].

There is also an international aspect to the FAA's failure. Historically, reciprocal agreements have been in place to allow airplanes to be approved faster. If one country approved of a model, it was typically accepted in other countries without delay. In light of the MAX disasters, however, European, Chinese, and Indian regulators have made it clear that they wish to approve planes independently of one another [27].

21.3.3 Recognition of the Need for Improved Training

Even after the FAA (and other national regulators) approves the plane, the 737 MAX planes will not be immediately ready to fly like they once were. Since the investigations began, Boeing has formally announced that pilots will be required to complete full-motion simulator training before being certified to fly the 737 MAX. Previously, Boeing pilots were only required to complete a "one-hour iPad lesson to fly the new 737 model." To complicate the new protocol, only 34 full-motion simulator training machines exist in the entire world. They are developed by two companies, the only two approved to make them. It would take years to train all of the 737 MAX pilots, even if the simulators were running 24/7. The Boeing 737 MAX will need to be rolled out gradually over time while training continues.

Note: In early 2021, Boeing reached a $2.5 billion settlement with the Department of Justice following its criminal investigation of the company. As reported in the *Wall Street Journal* [176], Boeing will pay $244 million in fines and $2.3 billion in compensation to customers and the families of the 346 people who perished in the two 737 MAX crashes. The FAA is still conducting a civil investigation of Boeing's activities, which could result in additional penalties. As stated in the article [176]:

Documents in the case reveal that for the first six months of the investigation, Boeing failed to cooperate with the grand jury probe and frustrated efforts by prosecutors delving into the matter. The filings also indicate that following the first MAX crash, one of the Boeing employees at the time misled FAA training experts, as well as some of the company's own officials, about why certain safety details were withheld from the FAA and MAX pilots before the agency's approval to carry passengers.

The FAA approved the Boeing 737 MAX's return to flight in November 2020, and they began to carry passengers in January 2021. Boeing added an alarm that alerts pilots to a mismatch of flight data from the two external sensors. It also rewired the system to prevent a short-circuit that leads to a crash if the pilot does not respond quickly. Perhaps most importantly, however, Boeing pilots are actually being trained on these new systems that they will be using [141].

NOTES

1. The similarities of the 737 MAX case to the Therac-25 – see the vignette presented in Chapter 17 – is striking. We have reuse of software systems from previous models. We also have the same focus on hardware by both the company and regulators, and the same overconfidence in software and its ability to account for potential hardware problems.
2. This is strikingly similar to the lack of documentation of errors in the Therac-25.
3. This is a common issue in medical devices/software also. Testing is often performed in major academic hospitals, where the physicians are highly experienced. How this testing translates to use in smaller settings is always an open problem.
4. This is a classic risk management failure. As reviewed in Chapter 5, risk is a combination of severity and probability. To control risk, one can either reduce the severity or the probability by appropriate measures. Here, the severity of the potential incident is clearly high, but reducing this would have required changes to the "hardware." The software solution chosen was assumed, erroneously, to be 100 percent foolproof (hence probability of harm = 0 percent), hence eliminating the risk in the minds of the engineers/designers. If that probability of harm was set to 1 percent, it would almost certainly have completely changed the process.
5. Please note that, at the time of this writing (2021), this is an ongoing story.

22 The Averted Y2K Crisis: Successful Crisis and Risk Management

Vignette written with Ellie Gabriel

INTRODUCTION

The changeover from 1999 to 2000 introduced a major risk of system failure in computer systems worldwide. This had to do with older software's use of two-digit numbers to store the year. As the year (in two digits) moved from 99 to 00, there were serious possibilities of many systems failing. This resulted in a huge effort to fix this problem across multiple major computer systems. In the end, despite minor issues, the change to the new millennium happened without any major problems.

22.1   The Y2K Problem

Y2K is the name given to the "Year 2000" problem. People were sure that the world was going to end after December 31, 1999 – see Figure 22.1 for a magazine cover from that period. At this point in time, most machines were operating based on systems that kept track of the year using two digits rather than four to save memory space during a time when memory was very expensive, leading to a potential confusion of the year 2000 with the year 1900, as both would be stored as "00".[1] Programmers erroneously believed that the band-aid solution they placed on the issue of accounting for dates in technology would solve itself through future upgrades. This could have catastrophic consequences in, for example, computing current bank balances, given the need to sort transactions by date, as a transaction on January 2, 2000 (01/02/00) would appear as older than one from December 20, 1999 (12/20/99). Hence a charge might be rejected as the balance in the account as of 01/02/00 might appear to be zero.

There was concern that the industrial world would come to a halt due to a global and indefinite power outage of all electronics once that 99 rolled over to 00 as the twenty-first century began. For machines that made predictions for the future based on the years or kept track of files based on expiration dates, this would present an especially detrimental circumstance. As Senator Chris Dodd, vice chairman of the US Senate's Special Committee on the Year 2000 Technology Problem, put it: "The question is not will there be disruptions, but how severe the disruptions will be" [260].

Figure 22.1 "End of the world" fears in the media. The January 18, 1999 cover of *TIME* magazine spurs fear about the threat of the Y2K bug. Image taken from Rothman [222].

22.1.1 A Technological World at Risk

In the late twentieth century, people began to directly depend on machines and computer systems infrastructure for critical tasks. Machines became part of daily life. This includes sectors such as banking, government systems, power plants, food sources, and transportation. The concern was widespread [104]. The international financial market, in particular, feared that the Y2K bug would prevent money withdrawals and transactions from taking place. Mainframe computers – those that are designed to deal with large inputs and outputs of data such as in banking, insurance, and supply chain – appeared most vulnerable. Besides computers, many devices containing computer chips (elevators, ATMs, temperature-control systems, medical equipment) were believed to be at risk; this made it essential to check and recheck these "embedded systems" for sensitivity, and hence vulnerability, to calendar dates [23].

While the costs associated with computer memory fell leading up to the 2000s, newer "better" computer systems had to be compatible with older versions to function – a computer that accepted four-digit dates needed to be able to interface with computers that accepted only two-digit dates. This required more active effort than expected, leading to the spread of the effects of the Y2K bug around the globe rather than containment. Technology running on software that recorded dates with two digits continued to be commonplace [260].

22.2 Preparations

22.2.1 Early Actions

The Y2K threat was not a last-minute realization. Computer scientists recognized the potential catastrophe in the 1990s (some even argued that this was already too late). They spent years preparing for the Y2K "bug" that they expected would set back all of the revolutionary advancements that were made in technology during the last part of the twentieth century. The US Senate established the Subcommittee on Financial Services and Technology in 1996 to study the problem and begin addressing Y2K compliance. In 1997 alone, the technology consulting firm Cap Gemini reported, "7% of a group of 128 large U.S. companies had experienced Y2K related problems. By March 1998, that number leaped to 37%" [260]. One research firm estimated that it would take up to $600 billion to fix the bug, and individual firms confirmed that they would likely individually lose up to $500 million from expected bug damages [104].

22.2.2 Legislative Steps in the United States

As a result of widespread public fears – see Figure 22.2 for a cartoon from the period – the US Congress passed *The Year 2000 Computer Remediation and Shareholder (CRASH) Protection Act of 1997* to require companies to disclose information related to their Y2K risk management plans. The following year, in 1998, a bill was passed to create a President's Council, which included officials from agencies such as the Federal Emergency Management Agency (FEMA), to prepare for the Y2K problem. FEMA, the government agency that is called upon for major natural disasters, was summoned to prepare the nation for a computer bug. It was felt that more than a mere task force was need to take on the complex Y2K threat [260].

In 1998, Cap Gemini America issued a Y2K preparedness survey to rank 13 economic sectors in the USA. The survey results showed that the US government was the least ready for Y2K, while the software industry was ranked highest in preparedness [23]. Partially in response to this, in 1999, *The Year 2000 Information and Readiness Disclosure Act* was passed to again encourage American businesses to share Y2K data by offering them limited liability protection for disclosing this, hence protecting them from legal action resulting from such disclosures.

Figure 22.2 Y2K "Backup" Plans. This joke gift is symbolic of how people without governmental or business power prepared for 2K on their own. Without technology, it would be back to traditional pencil and paper for everyone. Image from MacDonald [164].

22.2.3   International Concerns

Around the globe, other countries prepared for the crisis as well. In Europe, the British government readied its armed forces to supply assistance to local police in case critical systems like utilities, transportation, or emergency services failed. The European Union (EU) warned that efforts to solve the Y2K bug in Western Europe were inadequate, noting insufficiency in cross-border cooperation. Many Asian countries were already suffering from a continuous economic crisis at the time, making them, along with small and isolated countries, even less prepared than everyone else. Facing immense uncertainty but fearing global catastrophe, the United Nations (UN) convened in mid-December 1998 to discuss Y2K and to discuss a collaborative risk-management effort. This meeting also led to the establishment of the International Y2K Cooperation Center in Washington, DC [23].

22.2.4   The US Senate Special Committee Report of 1999

The committee issued an official investigative report in 1999 [260]. This begins by noting its frustration with the media's over- and understatement of the Y2K problem. Some news outlets enticed readers by claiming that Y2K was all a hoax and a conspiracy to grant more control to the government; others said that the world was doomed no matter what and advised people to pray for survival. In reality, the problem was major but definitely manageable through an organized group effort. The public first needed to look past their personal laptops and spreadsheets and think of the bigger picture – how their world was intertwined with layers of technology. The report stated that "the interdependent nature of technology systems makes the severity of possible disruptions difficult to predict." The report, which was published on February 24, 1999, noted that the Senate had seen progress since the inception of the Committee, but that it still feared that the country was lagging behind and would not be prepared for 2000 at the rate they were going [260].

The report expressed immediate concern about organizations that are critical to people's safety and well-being, such as doctors offices and small- and medium-sized businesses that were, by 1999, over 90 and 50 percent, respectively, non-Y2K compliant. Larger firms appeared more cognizant of the threat and were working to remedy it. To manage the entire country more effectively, the Special Committee held hearings to discuss nine critical economic sectors: utilities, healthcare, telecommunications, transportation, financial institutions, government, general business, litigation, and international affairs.

22.3   The Aftermath

22.3.1   Crisis Averted

The Senate report listed 10 potential Y2K solutions, including year interception (catch all date calculations and replace the incorrect ones), date–date expansion (convert

Best buy warns its Customers about Y2K. This picture comes from a blog post published by the electronics retailer Best Buy by John Vomhof Jr. that appeared on December 30, 1999. It was an attempt to warn its customers about the Y2K threat [264].

Figure 22.4 Buggy electronic sign illustrating Y2K bugs. An electronic sign at a French engineering university incorrectly displays the date on January 3, 2000 as January 3, 1900 as a result of a Y2K computer bug. There were some failures, but mostly these were insignificant. Image courtesy of Wikimedia Commons.

all two-digit dates into four-digit dates), and year shifting (use the 28-year calendar cycle to shift dates to the same century), among others. Ultimately, despite widespread concerns (e.g. see Figure 22.3), January 1, 2000 arrived with no major malfunctions.[2] The world, for the most part, smoothly transitioned into the new millennium – though some minor issues were observed, such as the one shown in Figure 22.4. The fact that the computer systems remained intact is a strong testament to the fact that the collective efforts to defeat the Y2K threat were successful. The threat was recognized early, significant effort was expended, and the whole process was followed through with an appropriate risk management plan. The Y2K compliance campaign that developed the computer system workarounds even ended up leading to technological improvements that stretched beyond 2000 [23].

The cause of the crisis can be attributed to (1) the human propensity to choose paths of least resistance and (2) the reluctance to overcome challenging issues. The Senate report noted that the core problem was a disconnect between the people who use technology and the people who develop it. Most people do not know how fragile information technology often is; however, the Y2K crisis, having been successfully addressed, may encourage leaders in all sectors of the economy to better understand our critical infrastructure. They must protect the parts of our world that keep it going around, especially as we become more reliant on them with time. The Committee hoped that [260]:

Y2K, as the first challenge of the information age, must leave a legacy of increased awareness and appreciation of information technology's role in social and economic advancement.

The success in containing the Y2K issue contains valuable lessons for risk management. The problems were identified early, senior leadership took the problems seriously, and significant resources were invested in mitigation strategies to resolve or contain the potential issues. It also shows the great role of collaboration during a global crisis, where politics were brushed aside as a result of the shared fear of the potential disaster that would have resulted had so many computer systems failed. Despite what some conspiracy-minded individuals may think, it is clear that 20 years ago we avoided a major disaster [159].

22.3.2   The Costs and the Y2K Blip

An estimated $300 billion was funneled into information technology in a major effort to create software patches and ways to work around the bug in the various systems that were affected. Additionally, the President's Council in the United States managed efforts made by private businesses to prepare their information systems for the new year [104]. It took a combined effort by businesses and government technology teams to check and fix the software before the turn of the century.

A secondary consequence was the Y2K blip – which was the result of excess inventory in many companies. While individual businesses could ensure that they could fix their own computer systems, they were concerned about other organizations up and down their supply chain. Essentially, businesses began stockpiling resources in case they became limited in production due to the Y2K bug in other industries. The Council of Economic Advisers (CEA) reported that Y2K preparatory spending "probably helps explain why real investment in computers and peripheral equipment in late 1998 was running more than 60% above its level a year earlier" [5].

NOTES

1. In the early 1960s, memory cost approximately $1/byte, whereas by 1999 the cost was reduced to approximately $1/million bytes.
2. One of the authors was an emergency contact person for a lab that New Year's Eve. Happily, he was able to stay home and enjoy the holiday.

References

[1] AAMI (Association for the Advancement of Medical Instrumentation). *AAMI TIR45:2012/(R)2018 Guidance on the use of AGILE practices in the development of medical device software*, August 20, 2012.

[2] AAMI (Association for the Advancement of Medical Instrumentation). *ANSI/AAMI/IEC 62304:2006/A1:2016 Medical device software – software life cycle processes. (Amendment)*, 2016.

[3] ABC123 (Anonymous). The failed launch of www.healthcare.gov. *HBS Digital Initiative*, November 18, 2016. https://digital.hbs.edu/platform-rctom/submission/the-failed-launch-of-www-healthcare-gov/.

[4] Acumen Research and Consulting. Digital health market value to reach USD 511 billion by 2026: the global digital health market size was valued at USD 95.7 billion in 2018 and expanding at a CAGR of 27.7% from 2019 to 2026, November 12, 2019. www.prnewswire.com/news-releases/digital-health-market-value-to-reach-usd-511-billion-by-2026-acumen-research-and-consulting-300956297.html.

[5] T. Aeppel. Manufacturers plan to set aside extra Inventory as Y2K safeguard. *Wall Street Journal*, February 9, 1999. www.wsj.com/articles/SB918519219843389000.

[6] N. Akhtar and A. Mian. Threat of adversarial attacks on deep learning in computer vision: a survey. *IEEE Access*, 6:14410–14430, 2018.

[7] T.A. Alspaugh. Software process models. www.thomasalspaugh.org/pub/fnd/softwareProcess.html.

[8] S. Amershi, A. Begel, C. Bird, DeLine, H. Gall, E. Kamar, N. Nagappan, B. Nushi, and T. Zimmermann. Software engineering for machine learning: a case study. In *Proceedings of the 41st International Conference on Software Engineering: Software Engineering in Practice*, ICSE-SEIP '19, pp. 291–300. IEEE Press, 2019. https://doi.org/10.1109/ICSE-SEIP.2019.00042.

[9] ANSI (American National Standards Institute) and The Santa Fe Group/Internet Security Alliance. The financial impact of breached protected health information: a business case for enhanced PHI security. Technical report, ANSI, 2012.

[10] Arterys FDA 510K Premarket Notification, May 2017. www.accessdata.fda.gov/scripts/cdrh/cfdocs/cfpmn/pmn.cfm?ID=K163253.

[11] Arterys Inc. Homepage. https://arterys.com/.

[12] ASCO CancerLinQ. Homepage. www.cancerlinq.org/.

[13] ASQ (American Society for Quality). Homepage. https://asq.org.

[14] J. Atherton. Development of the electronic health record. *Virtual Mentor*, 13(3):186–189, 2011. https://journalofethics.ama-assn.org/article/development-electronic-health-record/2011-03.

[15] B. Aulet. *Disciplined Enterpreneurship*. Wiley, 2013.

[16] J. Avorn. The $2.6 billion pill: methodologic and policy considerations. *N Engl J Med*, 372(20):1877–1879, 2015. www.ncbi.nlm.nih.gov/pubmed/25970049.

[17] K. Beck. *Extreme Programming Explained: Embrace Change*. Addison-Wesley, 2000.

[18] K. Beck, M. Beedle, A. van Bennekum, A. Cockburn, W. Cunningham, M. Fowler, J. Grenning, J. Highsmith, A. Hunt, R. Jeffries, J. Kern, B. Marick, R.C. Martin, S. Mellor, K. Schwaber, J. Sutherland, and D. Thomas. Manifesto for Agile software development, 2001. www.agilemanifesto.org/.

[19] S. Benjamens, P. Dhunnoo, and B. Meskó. The state of artificial intelligence-based FDA-approved medical devices and algorithms: an online database. *NPJ Digit Med*, 3:118, 2020. https://doi.org/10.1038/s41746-020-00324-0.

[20] J.M. Bland and D.G. Altman. Comparing methods of measurement: why plotting difference against standard method is misleading. *Lancet*, 346(8982):1085–1087, 1995.

[21] W. Boston. How Volkswagen's $50 billion plan to beat Tesla short-circuited: faulty software set back a bid by the world's largest car maker for electric-vehicle dominance. *Wall Street journal*, January 19, 2021. www.wsj.com/articles/how-volkswagens-50-billion-plan-to-beat-tesla-short-circuited-11611073974.

[22] BrainLAB. BrainLAB VectorVision cranial. www.brainlab.com/.

[23] Britannica. Y2K bug, May 18, 2020. www.britannica.com/technology/Y2K-bug.

[24] R. Buechi, L. Faes, L.M. Bachmann, M.A. Thiel, N.S. Bodmer, M.K. Schmid, O. Job, and K.R. Lienhard. Evidence assessing the diagnostic performance of medical smartphone apps: a systematic review and exploratory meta-analysis. *BMJ Open*, 7(12):e018280, 2017.

[25] L.A. Burke and A.M. Ryan. The complex relationship between cost and quality in US health care. *Virtual Mentor*, 16(2):124–130, 2014. https://journalofethics.ama-assn.org/article/complex-relationship-between-cost-and-quality-us-health-care/2014-02.

[26] R. Buys, J. Claes, D. Walsh, N. Cornelis, K. Moran, W. Budts, C. Woods, and V.A. Cornelissen. Cardiac patients show high interest in technology enabled cardiovascular rehabilitation. *BMC Med Inform Decis Mak*, 16(95), 2016. https://doi.org/10.1186/s12911-016-0329-9.

[27] D. Campbell. The ancient computers in the Boeing 737 MAX are holding up a fix. *The Verge*, April 9, 2020. www.theverge.com/2020/4/9/21197162/boeing-737-max-software-hardware-computer-fcc-crash.

[28] CBS News. The phishing email that hacked the account of John Podesta, October 28, 2016. www.cbsnews.com/news/the-phishing-email-that-hacked-the-account-of-john-podesta/.

[29] S. Chacon and B. Straub. *Pro Git*, 2nd ed. Apress, 2014.

[30] ChinaMed Device (CMD). Homepage. https://chinameddevice.com/.

[31] K. Chinzei, A. Shimizu, K. Mori, K. Harada, H. Takeda, M. Hashizume, M. Ishizuka, N. Kato, R. Kawamori, S. Kyo, K. Nagata, T. Yamane, I. Sakuma, K. Ohe, and M. Mitsuishi. Regulatory science on AI-based medical devices and systems. *Adv Biomed Eng*, 7:118–123, 2018.

[32] N. Chiu, A. Kramer, and A. Shah. 2020 midyear digital health market update: unprecedented funding in an unprecedented time, 2020. https://rockhealth.com/insights/2020-midyear-digital-health-market-update-unprecedented-funding-in-an-unprecedented-time/.

[33] J. Cleland-Huang. Don't fire the architect! Where were the requirements? *IEEE Software*, 31(2):27–29, 2014.

[34] C. Cochran. *ISO 9001:2015 in Plain English*. Paton Professional, 2015.

[35] J. Cohen. A coefficient of agreement for nominal scales. *Educ and Psychol Meas*, 20(1):37–46, 1960. https://doi.org/10.1177/001316446002000104.

[36] V. Combs. Why shortcuts lead to failure: lessons from app disaster in Iowa caucus. *Tech Republic*, February 6, 2020. www.techrepublic.com/article/why-shortcuts-lead-to-failure-lessons-from-app-disaster-in-iowa/.

[37] A. Cooper. *The Inmates Are Running the Asylum: Why High Tech Products Drive Us Crazy and How to Restore the Sanity*. Pearson Education, 2004.

[38] N. Corasaniti, S. Frenkel, and N. Perlroth. App used to tabulate votes is said to have been inadequately tested. *New York Times*, Feburary 3, 2020. www.nytimes.com/2020/02/03/us/politics/iowa-caucus-app.html.

[39] A. Coravos, J.C. Goldsack, D.R. Karlin, C. Nebeker, E. Perakslis, N. Zimmerman, and M.K. Erb. Digital medicine: a primer on measurement. *Digit Biomark*, 3:31–71, 2019. www.karger.com/Article/FullText/500413.

[40] CTTI (Clinical Trials Transformation Initiative). Homepage. www.ctti-clinicaltrials.org/.

[41] CTTI (Clinical Trials Transformation Initiative). Decentralized clinical trials. www.ctti-clinicaltrials.org/projects/decentralized-clinical-trials.

[42] CTTI (Clinical Trials Transformation Initiative). Digital health trials. www.ctti-clinicaltrials.org/projects/mobile-technologies.

[43] CTTI (Clinical Trials Transformation Initiative). Project: real-world data. www.ctti-clinicaltrials.org/projects/real-world-data.

[44] Decibio Insights. How consumerization is driving evolution of digital health through strategic partnerships, November 7, 2019.

[45] Deloitte Centre for Health Solutions. Measuring the return from pharmaceutical innovation, 2019. www2.deloitte.com/uk/en/pages/life-sciences-and-healthcare/articles/measuring-return-from-pharmaceutical-innovation.html.

[46] P. Densen. Challenges and opportunities facing medical education. *Trans Am Clin Climatol Assoc*, 122:48–58, 2011. www.ncbi.nlm.nih.gov/pubmed/21275727.

[47] Department of Health and Human Services (DHSS). CMS management of the federal marketplace, a case study, February 2016. https://oig.hhs.gov/oei/reports/oei-06-14-00350.pdf.

[48] Department of Health and Human Services (DHSS), Office of Civil Rights. Guidance regarding methods for de-identification of protected health information in accordance with the Health Insurance Portability and Accountability Act (HIPAA) Privacy Rule, November 26, 2012. www.hhs.gov/hipaa/for-professionals/privacy/special-topics/de-identification/index.html.

[49] Department of Health and Human Services (DHSS), Office of Civil Rights. HIPAA administrative simplification: regulation text, March 2013. www.hhs.gov/sites/default/files/hipaa-simplification-201303.pdf.

[50] Digital Therapeutics Alliance. Homepage. www.dtxalliance.org/.

[51] J.A. DiMasi, H.G. Grabowski, and R.W. Hansen. Innovation in the pharmaceutical industry: new estimates of R&D costs. *J Health Econ*, 47:20–33, 2016.

[52] DiMe (Digital Medicine Society). Homepage. www.dimesociety.org/.

[53] DIN (Deutsches Institut für Normung/German Institute for Standardization). *DIN SPEC 92001-1:2019-4 Artificial intelligence – life cycle processes and quality requirements – part 1: quality meta model*, April 2019.

[54] T. Doherty and T. Lindeman. The problems that led to the Boeing 737 MAX grounding. *Politico*, March 15, 2019. www.politico.com/story/2019/03/15/boeing-737-max-grounding-1223072.

[55] R. Duda and P. Hart. *Pattern Classification and Scene Analysis*. Wiley, 1973.

[56] Editorial Board. Boeing's fatal lessons. *Wall Street Journal*, December 17, 2019. www.wsj.com/articles/boeings-fatal-lessons-11576628330.

[57] A. Eklund, T.E. Nichols, and H. Knutsson. Cluster failure: why fMRI inferences for spatial extent have inflated false-positive rates. *Proc Nat Acad Sci*, 113(28):7900–7905, 2016.

[58] M. Enos. Fortune, Shadow Inc.: how a company with 120 Facebook likes ended up at the center of the Iowa Caucus firestorm. *Fortune*, February 6, 2020. https://fortune.com/2020/02/06/shadow-app-acronym-iowa-caucus-results/.

[59] Epic. Homepage. www.epic.com/.

[60] Epic. HL7v2. https://open.epic.com/Interface/HL7v2.

[61] A. Esteva, A. Robicquet, B. Ramsundar, V. Kuleshov, M. DePristo, K. Chou, C. Cui, G. Corrado, S. Thrun, and J. Dean. A guide to deep learning in healthcare. *Nat Med*, 25(1):24–29, 2019.

[62] EU Medical Device Coordination Group. MDCG 2019-11: guidance on qualification and classification of software in Regulation (EU) 2017/745 – MDR and Regulation (EU) 2017/746 – IVDR, October 2019. https://ec.europa.eu/health/system/files/2020-09/md_mdcg_2019_11_guidance_en_0.pdf

[63] European Medicines Agency. Questions and answers: qualification of digital technology-based methodologies to support approval of medicinal products, June 1, 2020. www.ema.europa.eu/en/documents/other/questions-answers-qualification-digital-technology-based-methodologies-support-approval-medicinal_en.pdf.

[64] European Union. *Regulation (EU) 2017/745 of the European Parliament and of the Council of 5 April 2017 on medical devices, amending Directive 2001/83/EC, Regulation (EC) No 178/2002 and Regulation (EC) No 1223/2009 and repealing Council Directives 90/385/EEC and 93/42/EEC, as amended*, April 5, 2017. https://eur-lex.europa.eu/legal-content/EN/TXT/PDF/?uri=CELEX:32017R0745.

[65] European Union. General Data Protection Regulation (GDPR), 2018. https://gdpr.eu.

[66] P. Farrugia, B.A. Petrisor, F. Farrokhyar, and M. Bhandari. Practical tips for surgical research: research questions, hypotheses and objectives. *Can J Surg*, 53(4):278–281, 2010.

[67] FDA. *FDA's* technology modernization action plan, September 18, 2019. www.fda.gov/about-fda/reports/fdas-technology-modernization-action-plan.

[68] FDA. FDA launches the Digital Health Center of Excellence, September 22, 2020. www.fda.gov/news-events/press-announcements/fda-launches-digital-health-center-excellence.

[69] FDA. Medical devices; current good manufacturing practice final rule; quality system regulation. *Federal Register*, 61(195), October 7, 1996. www.fda.gov/medical-devices/postmarket-requirements-devices/quality-system-qs-regulationmedical-device-good-manufacturing-practices.

[70] FDA. General principles of software validation; final guidance for industry and FDA staff, January 11, 2002. www.fda.gov/regulatory-information/search-fda-guidance-documents/general-principles-software-validation.

[71] FDA. Guidance for the content of premarket submissions for software contained in medical devices, May 11, 2005. www.fda.gov/regulatory-information/search-fda-guidance-documents/guidance-content-premarket-submissions-software-contained-medical-devices.

[72] FDA. Medical device reporting regulation history, March 22, 2018. www.fda.gov/medical-devices/mandatory-reporting-requirements-manufacturers-importers-and-device-user-facilities/medical-device-reporting-regulation-history.

[73] FDA. PMA labeling, September 27, 2018. www.fda.gov/medical-devices/premarket-approval-pma/pma-labeling.

[74] FDA. Regulatory controls, March 27, 2018. www.fda.gov/medical-devices/overview-device-regulation/regulatory-controls.

[75] FDA. Statistical assessment methodology for diagnostics and biomarkers, August 6, 2018. www.fda.gov/medical-devices/cdrh-research-programs/statistical-assessment-methodology-diagnostics-and-biomarkers.

[76] FDA. Clinical decision support software: draft guidance for industry and Food and Drug Administration staff, September 27, 2019. www.fda.gov/media/109618/download.

[77] FDA. De novo classification request, November 20, 2019. www.fda.gov/medical-devices/premarket-submissions/de-novo-classification-request.

[78] FDA. General wellness: policy for low risk devices. Guidance for industry and Food and Drug Administration staff, September 27, 2019. www.fda.gov/media/90652/download.

[79] FDA. Investigational device exemption (IDE), December 13, 2019. www.fda.gov/medical-devices/premarket-submissions-selecting-and-preparing-correct-submission/investigational-device-exemption-ide.

[80] FDA. Policy for device software functions and mobile medical applications. Guidance for industry and Food and Drug Administration staff, September 27, 2019. www.fda.gov/media/80958/download.

[81] FDA. Premarket Approval (PMA), May 16, 2019. www.fda.gov/medical-devices/premarket-submissions-selecting-and-preparing-correct-submission/premarket-approval-pma.

[82] FDA. Proposed regulatory framework for modifications to artificial intelligence/machine learning (AI/ML)-based software as a medical device (SaMD).

Discussion paper and request for feedback, April 2, 2019. www.fda.gov/media/122535/download.

[83] FDA. Regonized consensus standards 051,13-19, January 14, 2019. www.accessdata.fda.gov/scripts/cdrh/cfdocs/cfstandards/detail.cfm?standard_identification_no=38829.

[84] FDA. Clinical performance assessment: considerations for computer-assisted detection devices applied to radiology images and radiology device data in premarket notification (510(k)) submissions: Guidance for industry and FDA staff, January 22, 2020. www.fda.gov/media/77642/download.

[85] FDA. Digital Health Software Precertification (Pre-Cert) Program, September 14, 2020. www.fda.gov/medical-devices/digital-health/digital-health-software-precertification-pre-cert-program.

[86] FDA. Premarket Notification 510(k), March 13, 2020. www.fda.gov/medical-devices/premarket-submissions/premarket-notification-510k.

[87] FDA. Cybersecurity guidelines. www.fda.gov/medical-devices/digital-health-center-excellence/cybersecurity#guidance.

[88] FDA, Center for Devices and Radiological Health. Design control guidance for medical device manufacturers, March 11, 1997. www.fda.gov/media/116573/download.

[89] FDA, Center for Devices and Radiological Health. Applying human factors and usability engineering to medical devices: guidance for industry and Food and Drug Administration staff, February 3, 2016. www.fda.gov/regulatory-information/search-fda-guidance-documents/applying-human-factors-and-usability-engineering-medical-devices.

[90] FDA, Center for Devices and Radiological Health. Digital health innovation plan, 2020. www.fda.gov/media/106331/download.

[91] FDA, Center for Devices and Radiological Health. Artificial intelligence/machine learning (AI/ML)-based software as a medical device (SaMD) action plan, January 2021. www.fda.gov/media/145022/download.

[92] FDA, Center for Devices and Radiological Health. Software as medical device (SAMD): clinical evaluation. Guidance for industry and Food and Drug Administration staff, December 8, 2017. www.fda.gov/medical-devices/digital-health/software-medical-device-samd.

[93] Figma. Homepage. www.figma.com/.

[94] R.A. Fisher. *Statistical Methods for Research Workers*. Oliver & Boyd, 1925.

[95] K. Forsberg and H. Mooz. The relationship of systems engineering to the project cycle. In *First Annual Symposium of the National Council on Systems Engineering (NCOSE)*, October 1991.

[96] A. Gawande. Why doctors hate their computers. *The New Yorker*, November 5, 2018. www.newyorker.com/magazine/2018/11/12/why-doctors-hate-their-computers.

[97] D. Gladstone and L. Gladstone. *Venture Capital Investing: The Complete Handbook for Investing in Private Businesses for Outstanding Profits*. FT Prentice-Hall, 2004.

[98] E. Glazer, D. Seetharaman, and A. Corse. The shoestring app developer behind the Iowa caucus debacle. *Wall Street Journal*, February 5, 2020. www.wsj.com/articles/the-shoestring-app-developer-behind-the-iowa-caucus-debacle-11580904037.

[99] A. Goldstein. HHS failed to heed many warnings that HealthCare.gov was in trouble. *Washington Post*, February 23, 2016. http://wapo.st/1oZmeTF?tid=ss_tw.

[100] I. Goodfellow, Y. Bengio, and A. Courville. *Deep Learning*. MIT Press, 2016. www.deeplearningbook.org/.

[101] Google Patents. Homepage. https://patents.google.com/.

[102] Government of Japan. *Act on Securing Quality, Efficacy and Safety of Products Including Pharmaceuticals and Medical Devices*, 1969.

[103] W.E.L Grimson, R. Kikinis, F. Jolesz, and P. McL. Black. Image-guided surgery. *Scientific American*, 280(6):62–69, 1999.

[104] C. Halton. Y2K. *Investopedia*, March 18, 2020. www.investopedia.com/terms/y/y2k.asp.

[105] M. Hampson, D. Scheinost, M. Qiu, J. Bhawnani, C.M. Lacadie, J.F. Leckman, R.T. Constable, and X. Papademetris. Biofeedback of real-time functional magnetic resonance imaging data from the supplementary motor area reduces functional connectivity to subcortical regions. *Brain Connect*, 1(1):91–98, 2011.

[106] X. He and E. Frey. ROC, LROC, FROC, AFROC: an alphabet soup. *J Am Coll Radiol*, 6(9):652–655, 2009. https://doi.org/10.1016/j.jacr.2009.06.001.

[107] M.L. Head, L. Holman, R. Lanfear, A.T. Kahn, and M.D. Jennions. The extent and consequences of p-hacking in science. *PLoS Biology*, 13(3), 2015. www.ncbi.nlm.nih.gov/pmc/articles/PMC4359000.

[108] Heads of Medicines Agencies and European Medicines Agency. HMA-EMA Joint Big Data Taskforce – summary report, February 13, 2019. www.ema.europa.eu/en/documents/minutes/hma/ema-joint-task-force-big-data-summary-report_en.pdf.

[109] Health Canada. Software as a medical device (SaMD): definition and classification, December 18, 2019. www.canada.ca/en/health-canada/services/drugs-health-products/medical-devices/application-information/guidance-documents/software-medical-device-guidance-document.html.

[110] Health Level Seven International. Homepage. www.hl7.org/.

[111] R.M. Henig. The Dalkon Shield disaster. *Washington Post*, November 17, 1985. www.washingtonpost.com/archive/entertainment/books/1985/11/17/the-dalkon-shield-disaster/6c58f354-fa50-46e5-877a-10d96e1de610/.

[112] I. Hernandez, A. San-Juan-Rodriguez, C.B. Good, and W.F. Gellad. Changes in list prices, net prices, and discounts for branded drugs in the US, 2007–2018. *JAMA*, 323(9):854–862, 2020. www.ncbi.nlm.nih.gov/pubmed/30589626.

[113] How Stuff Works. How caucus work: the Iowa caucus. https://people.howstuffworks.com/question721.htm#pt1.

[114] H.K. Huang. Short history of PACS. Part I: USA. *Eur J Radiol*, 78(2):163–176, 2011.

[115] S.B. Hulley, S.R. Cummings, W.S. Browner, D.G. Grady, and T.B. Newman. *Designing Clinical Research*. Wolters Kluwer, 2013.

[116] P. Hunter. The big health data sale: as the trade of personal health and medical data expands, it becomes necessary to improve legal frameworks for protecting patient anonymity, handling consent and ensuring the quality of data. *EMBO Rep*, 17(8):1103–1105, 2016. www.ncbi.nlm.nih.gov/pubmed/27402546.

[117] P.S. Hussey, S. Wertheimer, and A. Mehrotra. The association between health care quality and cost: a systematic review. *Ann Internal Med*, 158(1):27–34, 2013. www.ncbi.nlm.nih.gov/pmc/articles/PMC4863949/.

[118] IDEO. *Human Centered Design Toolkit*, 2nd ed. IDEO, 2011.

[119] International Electrotechnical Commission (IEC). *IEC 62304 Medical device software – software life cycle processes*, May 2006.

[120] India Ministry of Health and Family Welfare. Medical device rules, January 31, 2017.

[121] Institute for Human Data Science (IQVIA). The growing value of digital health: evidence and impact on human health and the healthcare system, November 7, 2017. www.iqvia.com/insights/the-iqvia-institute/reports/the-growing-value-of-digital-health.

[122] International Electrotechnical Commission (IEC). *IEC 62366-1 Medical devices – part 1: application of usability engineering to medical devices*, February 2015.

[123] International Electrotechnical Commission (IEC). *IEC 62366-1 Medical devices – part 1: application of usability engineering to medical devices (Amendment 1)*, June 2020.

[124] International Electrotechnical Commission (IEC). *IEC 62366-2 Medical devices – part 2: guidance on the application of usability engineering to medical devices*, April 2016.

[125] International Medical Devices Regulator Forum (IMDRF), SaMD Working Group. Software as a medical device (SaMD): key definitions, December 9, 2013. IMDRF/SaMD WG/N10. www.imdrf.org/docs/imdrf/final/technical/imdrf-tech-131209-samd-key-definitions-140901.pdf.

[126] International Medical Devices Regulator Forum (IMDRF), SaMD Working Group. Software as medical device: possible framework for risk categorization and corresponding considerations, September 18, 2014. www.imdrf.org/sites/default/files/docs/imdrf/final/technical/imdrf-tech-140918-samd-framework-risk-categorization-141013.pdf.

[127] International Medical Devices Regulator Forum (IMDRF), SaMD Working Group. Software as a medical device (SaMD): application of quality management system, October 2, 2015. www.imdrf.org/docs/imdrf/final/technical/imdrf-tech-151002-samd-qms.pdf.

[128] International Medical Devices Regulator Forum (IMDRF), Management Committee. Statement regarding use of IEC 62304:2006 "Medical device software – software life cycle processes", October 2, 2015. www.imdrf.org/docs/imdrf/final/procedural/imdrf-proc-151002-medical-device-software-n35.pdf.

[129] International Medical Devices Regulator Forum (IMDRF), Medical Device Cybersecurity Working Group. Principles and practices for medical device cybersecurity, March 18, 2020. www.imdrf.org/docs/imdrf/final/technical/imdrf-tech-200318-pp-mdc-n60.pdf.

[130] International Medical Devices Regulator Forum (IMDRF), Medical Device Clinical Evaluation Working Group. Clinical investigation, October 10, 2019. www.imdrf.org/sites/default/files/docs/imdrf/final/technical/imdrf-tech-191010-mdce-n57.pdf.

[131] International Standards Organization (ISO). *ISO:9001 Quality management systems*, 5th ed. 2015.

[132] International Standards Organization (ISO). *ISO:13485 Medical devices – quality management systems – requirements for regulatory purposes*, 2016.

[133] International Standards Organization (ISO). *ISO:14971:2019 Medical devices – application of risk management to medical devices*, 2019.

[134] International Standards Organization (ISO). *ISO/IEC/IEEE:90003 Software engineering – guidelines for the application of ISO 9001:2015 to computer software*, 2018.

[135] International Standards Organization (ISO). Friendship among equals: recollections from ISO's first fifty years, 1997. www.iso.org/files/live/sites/isoorg/files/about%20ISO/docs/en/Friendship_among_equals.pdf.

[136] F. Isensee, P.F. Jaeger, S.A. Kohl, J. Petersen, and K.H. Maier-Hein. nnU-Net: a self-configuring method for deep learning-based biomedical image segmentation. *Nat Methods*, 18:203–211, 2021

[137] Janssen. Janssen leverages wearable technology to reimagine clinical trial design, November 16, 2019. www.prnewswire.com/news-releases/janssen-leverages-wearable-technology-to-reimagine-clinical-trial-design-300959419.html.

[138] A. Jarvis and P. Palmes. *ISO 9001:2015: Understand, Implement, Succeed!* Addison-Wesley, 2015.

[139] Johnson & Johnson. Johnson & Johnson launches Heartline the first-of-its-kind, virtual study designed to explore if a new iPhone app and Apple Watch can help reduce the risk of stroke, Feburary 25, 2020. www.prnewswire.com/news-releases/johnson--johnson-launches-heartline-the-first-of-its-kind-virtual-study-designed-to-explore-if-a-new-iphone-app-and-apple-watch-can-help-reduce-the-risk-of-stroke-301010792.html.

[140] Johner Institute North America. Homepage. https://johner-institute.com/.

[141] E.M. Johnson. Factbox: key changes to Boeing's 737 MAX after fatal crashes. *Reuters*, November 17, 2020. www.reuters.com/article/us-boeing-737max-changes-factbox/factbox-key-changes-to-boeings-737-max-after-fatal-crashes-idUSKBN27X33M.

[142] N. Kano, S. Nobuhiku, T. Fumio, and T. Shinichi. Attractive quality and must-be quality. *J Japan Soc Quality Contr. (in Japanese)*, 14(2):39–48, 1984.

[143] B. Kappe. *Accelerating Medical Product Development: Applying Agile Methods to Shorten Timelines, Reduce Risk and Improve Quality*. Orthogonal, 2020. http://orthogonal.io/insights/.

[144] D. Kelley and T. Kelley. *Creative Confidence: Unleashing the Creative Potential within Us All*. William Collins, 2013.

[145] D. Kidder and H. Hindi. *The Startup Playbook: Secrets of the Fastest-Growing Startups from Their Founding Entrepreneurs*. Chronicle Books, 2012.

[146] J. Koebler and E. Mailberg. Here's the Shadow Inc. app that failed in Iowa last night. *Vice*, Feburary 4, 2020. www.vice.com/en/article/y3m33x/heres-the-shadow-inc-app-that-failed-in-iowa-last-night.

[147] G. Kolata. The sad legacy of the Dalkon Shield. *New York Times*, December 6, 1987. www.nytimes.com/1987/12/06/magazine/the-sad-legacy-of-the-dalkon-shield.html.

[148] D. Kopf. The Guinness brewer who revolutionized statistics. https://priceonomics.com/the-guinness-brewer-who-revolutionized-statistics/.

[149] Korean Software Testing Qualifications Board and Chinese Software Testing Qualifications Board. Certified tester: AI testing – testing AI-based systems (AIT – TAI). Foundation level syllabus, 2019.

[150] G. Kurian, editor. *A Historical Guide to the U.S. Government*. Oxford University Press, 1988.

[151] J.J. Laffont and J. Tirole. The politics of government decision-making: a theory of regulatory capture. *Q J Econ*, 106(4):1089–1127, 1991.

[152] Y. LeCun, Y. Bengio, and G. Hinton. Deep learning. *Nature*, 521(7553):436–444, 2015.

[153] G. Lee and J. Brumer. Managing mission-critical government software projects: lessons learned from the HealthCare.gov project, 2017. www.businessofgovernment.org/ sites/default/files/Viewpoints%20Dr%20Gwanhoo%20Lee.pdf.

[154] N.G. Leveson. *Safeware: System Safety and Computers*. Addison-Wesley, 1995.

[155] N.G. Leveson. The Therac-25: 30 years later. *Computer*, 50(11):8–11, 2017.

[156] N.G. Leveson and C.S. Turner. An investigation of the Therac-25 accidents. *Computer*, 26(7):18–41, 1993.

[157] C. Linnane. Teladoc–Livongo $18.5 billion merger is a huge step forward for digital health, analysts say. *MarketWatch*, August 6, 2020. www.marketwatch.com/story/ teladoc-livongo-185-billion-merger-is-a-huge-step-forward-for-digital-health-analysts-say-2020-08-05.

[158] B. Littlewood and L. Strigini. The risks of software. *Scientific American*, 267(5):62–75, 1992

[159] Z. Loeb. The lessons of Y2K, 20 years later: Y2K became a punchline, but 20 years ago we averted disaster. *Washington Post*, December 30, 2019. www.washingtonpost .com/outlook/2019/12/30/lessons-yk-years-later/.

[160] S. Lohr. For big-data scientists, "janitor work" is key hurdle to insights. *New York Times*, August 2014. www.nytimes.com/2014/08/18/technology/for-big-data-scientists-hurdle-to-insights-is-janitor-work.html.

[161] E.N. Lorenz. *The Essence of Chaos (Jessie and John Danz Lectures)*. University of Washington Press, 1995.

[162] A. Loten. Testing could have prevented Iowa caucus App failure, experts say. *Wall Street Journal*, February 4, 2020. www.wsj.com/articles/testing-could-have-prevented-iowa-caucus-app-failure-experts-say-11580856659.

[163] J. Lynch. The worst computer bugs in history: rapid unanticipated disassembly of the Mars Climate Orbiter. *bugsnug*, September 12, 2017. www.bugsnag.com/blog/bug-day-mars-climate-orbiter.

[164] N. MacDonald. Y2K 20 years later: an oral history of the final moments of the 20th century at Seattle City Light, December 20, 2019. https://powerlines.seattle.gov/2019/ 12/20/y2k/.

[165] V. Martindale and A. Menache. The PIP scandal: an analysis of the process of quality control that failed to safeguard women from the health risks. *J R Soc Med*, 106(5):173–177, 2013.

[166] Mayo Clinic. Radiation therapy. www.mayoclinic.org/tests-procedures/radiation-therapy/ about/pac-20385162.

[167] E. McCallister, T. Grance, and K. Scarfone. *National Institute of Standards and Technology (NIST) Special Publication 800-122: Guide to Protecting the Confiden-*

tiality of Personally Identifiable Information (PII). NIST, 2020. https://nvlpubs.nist.gov/nistpubs/Legacy/SP/nistspecialpublication800-122.pdf.

[168] J. McCarthy. What is artificial inteligence?, November 12, 2007. http://jmc.stanford.edu/articles/whatisai/whatisai.pdf.

[169] J.J. McGough and S.V. Faraone. Estimating the size of treatment effects: moving beyond p values. *Psychiatry*, 6(10):21–29, 2009.

[170] M. Meadows. Promoting safe and effective drugs for 100 years. *FDA Consumer magazine: The Centennial Edition*, January–February 2006. www.fda.gov/about-fda/histories-product-regulation/promoting-safe-effective-drugs-100-years.

[171] Medical Imaging and Technology Alliance. DICOM: Digital Imaging and Communications in Medicine. www.dicomstandard.org/current/.

[172] Medical Imaging and Technology Alliance. DICOMWeb: DICOM standard for web-based medical imaging. www.dicomstandard.org/dicomweb/.

[173] Medicines & Healthcare Products Regulatory Agency (MHRA). Human factors and usability engineering: guidance for medical Devices including drug–device combination products, September 2017. https://assets.publishing.service.gov.uk/government/uploads/system/uploads/attachment_data/file/970563/Human-Factors_Medical-Devices_v2.0.pdf.

[174] Medicines & Healthcare Products Regulatory Agency (MHRA). Guidance: medical device stand-alone software including apps (including IVDMDs), November 2020. https://assets.publishing.service.gov.uk/government/uploads/system/uploads/attachment_data/file/999908/Software_flow_chart_Ed_1-08b-IVD.pdf.

[175] Medtronic Surgical Navigation Technologies. Medtronic StealthStation surgical navigation system. www.medtronic.com/us-en/healthcare-professionals/products/neurological/surgical-navigation-systems/stealthstation.html.

[176] D. Michaels, A. Tangel, and A. Pasztor. Boeing reaches $2.5 billion settlement of U.S. probe into 737 MAX crashes: agreement with Justice Department allows aerospace giant to avoid prosecution. *Wall Street Journal*, January 7, 2021. www.wsj.com/articles/boeing-reaches-2-5-billion-settlement-of-u-s-probe-into-737-max-crashes-11610054729.

[177] Microsoft. Threat modeling. www.microsoft.com/en-us/securityengineering/sdl/threatmodeling.

[178] Mishap Investigation Board. Phase I Report, Mars Climate Orbiter, November 1999. https://llis.nasa.gov/llis_lib/pdf/1009464main1_0641-mr.pdf.

[179] MONAI, Medical Open Network for AI. Homepage. https://monai.io/.

[180] B. Montgomery. Closed loop systems: future treatment for diabetes?, June 3, 2020. www.thediabetescouncil.com/closed-loop-systems-future-treatment-for-diabetes/.

[181] E. Morath, J. Hilsenrath, and S. Chaney. Record rise in unemployment claims halts historic run of job growth. *Wall Street Journal*, March 26, 2020. www.wsj.com/articles/the-long-run-of-american-job-growth-has-ended-11585215000.

[182] T. Moynihan. Samsung finally reveals why the Note 7 kept exploding. *Wired*, January 22, 2017. www.wired.com/2017/01/why-the-samsung-galaxy-note-7-kept-exploding/.

[183] R. Mullin. Tufts study finds big rise in cost of drug development. *Chemical and Engineering News*, November 20, 2014. https://cen.acs.org/articles/92/web/2014/11/Tufts-Study-Finds-Big-Rise.html.

[184] National Aeronautics and Space Administration (NASA). Mars mission, press kit, December 1998. www.jpl.nasa.gov/news/press_kits/mars98launch.pdf.

[185] National Cancer Institute. External beam radiation therapy to treat cancer. www.cancer.gov/about-cancer/treatment/types/radiation-therapy/external-beam.

[186] National Cancer Institute. Radiation therapy to treat cancer. www.cancer.gov/about-cancer/treatment/types/radiation-therapy.

[187] National Institute of Standards and Technology (NIST). Framework for improving critical infrastructure cybersecurity, April 16, 2018. www.nist.gov/cyberframework.

[188] National Medical Products Administration (NMPA). *Rules for classification of medical devices: (Decree No. 15 of China Food and Drug Administration)*, November 10, 2019. http://subsites.chinadaily.com.cn/nmpa/2019-10/11/c_415411.htm.

[189] National Medical Products Administration (NMPA). Technical guideline on AI-aided software, June 2019.

[190] Novartis. The Novartis biome: a catalyst for impactful digital collaboration. www.novartis.com/our-science/novartis-biome.

[191] R. Nunn, J. Parsons, and J. Shambaugh. A dozen facts about the economics of the U.S. health-care system, March 2020. www.brookings.edu/research/a-dozen-facts-about-the-economics-of-the-u-s-health-care-system/.

[192] Nuremberg Code. Trials of war criminals before the Nuremberg military tribunals under Control Council Law No. 10, 1949. https://history.nih.gov/display/history/Nuremberg+Code.

[193] J. Oberg. Why the Mars probe went off course [accident investigation]. *IEEE Spectrum*, 36(12):34–39, 1999.

[194] OECD (Organization for Economic Co-operation and Development). Improving healthcare quality in Europe: characteristics, effectiveness and implementation of different strategies, October 17, 2019. www.oecd.org/els/improving-healthcare-quality-in-europe-b11a6e8f-en.htm.

[195] OECD (Organization for Economic Co-operation and Development) and the European Union. *Health at a Glance: Europe 2016*. OECD, 2016. www.oecd-ilibrary.org/content/publication/9789264265592-en.

[196] OFFIS e.V. DCMTK: Dicom toolkit. https://dicom.offis.de/dcmtk.php.en.

[197] Otsuka. Otsuka and Proteus announce the first U.S. FDA approval of a digital medicine system: ABILIFY MYCITE (aripiprazole tablets with sensor). Press release, November 14, 2017.

[198] X. Papademetris, C. DeLorenzo, S. Flossmann, M. Neff, K.P. Vives, D.D. Spencer, L.H. Staib, and J.S. Duncan. From medical image computing to computer-aided intervention: development of a research interface for image-guided navigation. *Int J Med Robot*, 5(2):147–157, 2009.

[199] N. Papernot, P. McDaniel, S. Jha, M. Fredrikson, Z.B. Celik, and A. Swami. The limitations of deep learning in adversarial settings. In *2016 IEEE European Symposium on Security and Privacy (EuroS P)*, pp. 372–387, 2016.

[200] A. Papoulis and S.U. Pillai. *Probability, Random Variables, and Stochastic Processes*, 4th ed. McGraw Hill, 2002.

[201] A. Pasztor. Boeing finds new software problem that could complicate 737 MAX's return. *Wall Street Journal*, January 17, 2020. www.wsj.com/articles/boeing-finds-new-software-problem-that-could-complicate-737-max-return-11579290347.

[202] A. Pasztor. Congressional report faults Boeing on MAX design, FAA for lax oversight. *Wall Street Journal*, March 6, 2020. www.wsj.com/articles/congressional-report-says-max-crashes-stemmed-from-boeings-design-failures-and-lax-faa-oversight-11583519145.

[203] T.M. Pawlik and J.A. Sosa, editors. *Clinical Trials*, 2nd ed. Springer, 2020.

[204] Pear Therapeutics. Pear obtains FDA clearance of the first digital therapeutic to treat disease. Press release, September 14, 2017. https://peartherapeutics.com/fda-obtains-fda-clearance-first-prescription-digital-therapeutic-treat-disease/.

[205] J. Pearl and D. Mackenzie. *The Book of Why: The New Science of Cause and Effect*. Basic Books, 2018.

[206] PERFORCE. Helix RM. www.perforce.com/products/helix-requirements-management.

[207] Pharmaceutical Group of the European Union. PGEU statement: eHealth solutions in European Community pharmacies, November 2016. www.pgeu.eu/wp-content/uploads/2019/07/161102E-PGEU-Statement-on-eHealth-Final.pdf.

[208] O.S. Pianykh. *Digital Imaging and Communications in Medicine (DICOM): A Practical Introduction and Survival Guide*, 2nd ed. Springer, 2012.

[209] J.J. Plecs and J.H. Cochrane. Imagine what we could cure if draconian privacy regulations didn't keep key data from medical researchers. *Wall Street Journal*, November 25, 2018. www.wsj.com/articles/imagine-what-we-could-cure-1543176157.

[210] PyTorch. Homepage. https://pytorch.org/.

[211] Radiopaedia. Homepage. www.radiopaedia.com.

[212] R. Rasmussen, T. Hughes, J.R. Jenks, and J. Skach. Adopting Agile in an FDA regulated environment. In *Agile Conference*, pp. 151–155, 2009.

[213] F. Redmill and J. Rajan. *Human Factors in Safety-Critical Systems*. Butterworth-Heinemann, 1997.

[214] Republic of Korea Ministry of Food and Drug Safety. Regulations. www.mfds.go.kr/eng/brd/m_40/list.do.

[215] Republic of Korea Ministry of Food and Drug Safety, Medical Device Evaluation Department. Guideline on review and approval of artificial intelligence (AI) and big data-based medical devices (for industry), November 4, 2020. www.mfds.go.kr/eng/brd/m_40/down.do?brd_id=eng0011&seq=72623&data_tp=A&file_seq=1.

[216] S. Robertson and J. Robertson. *Mastering the Requirements Process: Getting Requirements Right*, 3rd ed. Addison-Wesley, 2012.

[217] O. Ronneberger, P. Fischer, and T. Brox. U-Net: convolutional networks for biomedical image segmentation. In N. Navab, J. Hornegger, W. Wells, and A. Frangi, editors, *Medical Image Computing and Computer-Assisted Intervention – MICCAI*. Springer, 2015. https://doi.org/10.1007/978-3-319-24574-4_28.

[218] Y. Ronquillo, A. Meyers, and S.J. Korvek. Digital health. *StatPearls (Internet)*, July 4, 2020. www.ncbi.nlm.nih.gov/books/NBK470260/.

[219] B.W. Rose. Fatal dose: radiation deaths linked to AECL computer errors, 1994. www.ccnr.org/fatal_dose.html.

[220] D.S. Rose. *Angel Investing: The Gust Guide to Making Money and Having Fun Investing in Startups.* Wiley, 2014.

[221] L. Rosen. *Open Source Licensing: Software Freedom and Intellectual Property Law.* Prentice Hall, 2004.

[222] L. Rothman. Remember Y2K? Here's how we prepped for the non-disaster. *TIME*, December 31, 2014. https://time.com/3645828/y2k-look-back/.

[223] W. Royce. Managing the development of large software systems: concepts and techniques. In *IEEE WESCON*, pp. 1–9, 1970. https://leadinganswers.typepad.com/leading_answers/files/original_waterfall_paper_winston_royce.pdf.

[224] W. Royce. TRW's Ada Process Model for incremental development of large software systems. In *Proceedings of the 12th International Conference on Software Engineering*, pp. 2–11, 1990.

[225] K.S. Rubin. *Essential Scrum: A Practical Guide to the Most Popular Agile Process.* Addison-Wesley, 2012.

[226] C. Rudin. Stop explaining black box machine learning models for high stakes decisions and use interpretable models instead. *Nat Machine Intell*, 1(5):206–215, 2019.

[227] T. Russotto and B.T. Wilson. Who's to blame for Boeing's 737 MAX saga? *Wall Street Journal*, December 29, 2019. www.wsj.com/articles/congressional-report-says-max-crashes-stemmed-from-boeings-design-failures-and-lax-faa-oversight-11583519145.

[228] B. Sahiner, B. Friedman, C. Linville, C. Ipach, E. Montgomery, E. Steinle Alexander, et al. Perspectives and good practices for AI and continuously learning systems in healthcare, 2018. www.exhibit.xavier.edu/health_services_administration_faculty/21/.

[229] R. Salay and K. Czarnecki. Using machine learning safely in automotive software: an assessment and adaption of software process requirements in ISO 26262. arXiv:1808.01614 [cs.LG], 2018. http://arxiv.org/abs/1808.01614.

[230] E. Sauerwein, F. Bailom, K. Matzler, and H.H. Hinterhuber. The Kano model: how to delight your customers. In *International Working Seminar on Production Economics*, volume 1, pp. 313–327, 1996.

[231] D. Scheinost, M. Hampson, M. Qiu, J. Bhawnani, R.T. Constable, and X. Papademetris. A graphics processing unit accelerated motion correction algorithm and modular system for real-time fMRI. *Neuroinformatics*, 11(3):291–300, 2013.

[232] R. Siegel. What if healthcare could start with technology? – Bernard Tyson, CEO Kaiser Permanente. *The Industrialist's Dilemma*, February 11, 2016. https://medium.com/the-industrialist-s-dilemma/what-if-healthcare-could-start-with-technology-bernard-tyson-ceo-kaiser-permanente-5052658a6212.

[233] D. Simas. Why we passed the Affordable Care Act in the first place? *The White House Archives: President Barack Obama*, October 30, 2013. https://obamawhitehouse.archives.gov/blog/2013/10/30/why-we-passed-affordable-care-act-first-place.

[234] Singapore Health Sciences Authority (HSA). *Regulatory guidelines for software medical devices: a lifecycle approach*, December 2019. www.hsa.gov.sg/docs/default-source/announcements/regulatory-updates/regulatory-guidelines-for-software-medical-devices–a-lifecycle-approach.pdf.

[235] Slicer. 3D-Slicer: self-test module. www.slicer.org/wiki/Documentation/Nightly/Developers/Tutorials/SelfTestModule.

[236] P. Spence. When the human body is the biggest data platform, who will capture value? *EY*, May 17, 2018. www.ey.com/en_us/digital/when-the-human-body-is-the-biggest-data-platform-who-will-capture-value.

[237] R. Spronk. Ringholm whitepaper: HL7 message examples: version 2 and version 3, November 16, 2007. http://ringholm.com/docs/04300_en.htm.

[238] Stanford Medicine News Center. Through Apple heart study, Stanford Medicine researchers show wearable technology can help detect atrial fibrillation, November 13, 2019. https://med.stanford.edu/news/all-news/2019/11/through-apple-heart-study--stanford-medicine-researchers-show-we.html.

[239] I. Stanley-Becker and M. Scherer. Iowa Democrats kept their App secret to prevent hacks. Instead, they got confusion and chaos. *Washington Post*, Feburary 4, 2020. www.washingtonpost.com/politics/iowa-democrats-kept-their-app-secret-to-prevent-hacks-instead-they-got-confusion-and-chaos/2020/02/04/fbd99654-4784-11ea-bc78-8a18f7afcee7_story.html.

[240] N. Statt. Apple confirms cloud gaming services like xCloud and Stadia violate App Store guidelines: new cloud gaming services from Google and Microsoft won't work on iOS. *The Verge*, August 6, 2020. www.theverge.com/2020/8/6/21357771/apple-cloud-gaming-microsoft-xcloud-google-stadia-ios-app-store-guidelines-violations.

[241] N. Statt. The App that broke the Iowa Caucuses was sent out through beta testing platforms. *The Verge*, February 4, 2020. www.theverge.com/2020/2/4/21122737/iowa-democractic-caucus-voting-app-android-testfairy-screenshots-app-store.

[242] S. Stern. Pillpack. MIT case study, MIT Sloan School of Management (2013).

[243] J.P. Swann. FDA's origin, February 1, 2018. www.fda.gov/about-fda/fdas-evolving-regulatory-powers/fdas-origin.

[244] C. Szegedy, W. Zaremba, I. Sutskever, J. Bruna, D. Erhan, I.J. Goodfellow, and R. Fergus. Intriguing properties of neural networks. In Yoshua Bengio and Yann LeCun, editors, *2nd International Conference on Learning Representations*, ICLR 2014, Banff, AB, Canada, April 14-16, 2014.

[245] H. Takeuchi and I. Nonaka. The new new product development game. *Harvard Business Review*, January 1986. https://hbr.org/1986/01/the-new-new-product-development-game.

[246] N.N. Taleb. *The Black Swan*, 2nd ed. Random House, 2010.

[247] N.N. Taleb. *Statistical Consequences of Fat Tails: Real World Preasymptotics, Epistemology, and Applications*. STEM Academic Press, 2020.

[248] A. Tangel, A. Pasztor, and M. Maremont. The four-second catastrophe: how Boeing doomed the 737 MAX. *Wall Street Journal*, August 16, 2019. www.wsj.com/articles/the-four-second-catastrophe-how-boeing-doomed-the-737-max-11565966629.

[249] Teladoc Health Inc. Teladoc Health reports first quarter 2019 results, April 30, 2019. www.globenewswire.com/news-release/2019/04/30/1813070/0/en/Teladoc-Health-Reports-First-Quarter-2019-Results.html.

[250] TensorFlow. Homepage. www.tensorflow.org/.

[251] M. Terry. The median cost of bringing a drug to market is $985 million, according to new study. *BioSpace.com*, March 4, 2020. www.biospace.com/article/median-cost-of-bringing-a-new-drug-to-market-985-million/.

[252] Therapeutic Goods Administration (TGA), Department of Health, Australian Government. Actual and potential harm caused by medical software: a rapid literature review of safety and performance issues, July 2020. www.tga.gov.au/resource/actual-and-potential-harm-caused-medical-software.

[253] D. Thompson. Health care just became the U.S.'s largest employer. *The Atlantic*, January 9, 2018. www.theatlantic.com/business/archive/2018/01/health-care-america-jobs/550079/.

[254] E. Tjoa and C. Guan. A survey on explainable artificial intelligence (XAI): towards medical XAI. arXiv:1907.07374, 2019. http://arxiv.org/abs/1907.07374.

[255] E. Topol. *Deep Medicine: How Artificial Intelligence Can Make Healthcare Human Again*. Basic Books, 2019.

[256] E. Topol. The Topol review: an independent report on behalf of the Secretary of State for Health and Social Care. NHS, Health Education England, February 2019. https://topol.hee.nhs.uk/wp-content/uploads/HEE-Topol-Review-2019.pdf.

[257] G. Travis. How the Boeing 737 MAX disaster looks to a software developer. *IEEE Spectrum*, April 18, 2019. https://spectrum.ieee.org/aerospace/aviation/how-the-boeing-737-max-disaster-looks-to-a-software-developer.

[258] F. Trotter and D. Uhlman. *Hacking Healthcare: A Guide to Standards, Workflows, and Meaningful Use*. O'Reilly Media, 2011.

[259] T.C. Tsai, E.J. Orav, and K.E. Joynt. Disparities in surgical 30-day readmission rates for Medicare beneficiaries by race and site of care. *Ann Surg*, 259(6):1086–1090, 2014. www.ncbi.nlm.nih.gov/pubmed/16432363.

[260] US Senate, Special Committee on the Year 2000 Technology Problem. Investigating the impact of the year 2000 problem, February 24, 1999. www.govinfo.gov/content/pkg/GPO-CPRT-106sprt10/pdf/GPO-CPRT-106sprt10.pdf.

[261] D.A. Vogel. *Medical Device Software Verification, Validation, and Compliance*. Artech House, 2011.

[262] Voluntis. Voluntis announces marketing authorization for Oleena, first digital therapeutic in oncology. Press release, July 31, 2019. www.voluntis.com/voluntis-announces-market-authorization-for-oleena-first-digital-therapeutic-in-oncology/.

[263] D. Volz, T. Parti, A. Corse, and R. McMillan. Iowa's tally-by-app experiment fails. *Wall Street Journal*, February 4, 2020. www.wsj.com/articles/iowa-caucus-results-delayed-by-apparent-app-issue-11580801699.

[264] J. Vomhof Jr. Geek Squad agents reflect on 20th anniversary of Y2K, December 30, 2019. https://corporate.bestbuy.com/geek-squad-agents-reflect-on-20th-anniversary-of-y2k/.

[265] J. Weiner, C. Marks, and M. Pauly. Effects of the ACA on health care cost containment. Issue Brief, Leondard Davis Institute of Health Economics, University of Pennsylvania, March 2, 2017. https://ldi.upenn.edu/brief/effects-aca-health-care-cost-containment.

[266] R.C. Welsh, J.E. Hardee, and S. Peltier. Slice-time correction in resting-state (and fMRI) gone bad. In *Organization of Human Brain Mapping Annual Meeting*, June 2014.

[267] L. Williams, E.M. Maximilien, and M. Vouk. Test-driven development as a defect-reduction practice. In *14th International Symposium on Software Reliability Engineering*, pp. 34–45, 2003.

[268] C.M. Wilson, D.P. Schreiber, J.D. Russell, and P. Hitchcock. Electron beam versus photon beam radiation therapy for the treatment of orbital lymphoid tumors. *Med Dosim*, 17(3):161–165, 1992.

[269] A. Wirth, C. Gates, and J. Smith. *Medical Device Cybersecurity for Engineers and Manufacturers*. Artech House, 2020.

[270] World Medical Association. Declaration of Helsinki: ethical principles for medical research involving human subjects, 1964–2013. www.wma.net/wp-content/uploads/2016/11/DoH-Oct2013-JAMA.pdf.

[271] P. Workman, G.F. Draetta, J.M. Schellens, and R. Bernards. How much longer will we put up with $100,000 cancer drugs? *Cell*, 168(4):579–583, 2017. https://pubmed.ncbi.nlm.nih.gov/28187281/.

[272] World Health Organization (WHO), Organization for Economic Co-operation and Development (OECD), and World Bank. Delivering quality health services: a global imperative for universal health coverage, July 5, 2018. www.worldbank.org/en/topic/universalhealthcoverage/publication/delivering-quality-health-services-a-global-imperative-for-universal-health-coverage.

[273] World Health Organization (WHO), Noncommunicable Diseases and Mental Health. Innovative care for chronic conditions: building blocks for action, 2002. www.who.int/chp/knowledge/publications/icccreport/en/.

[274] E. Wu, K. Wu, R. Daneshjou, D. Ouyang, D.E. Ho, and J. Zou. How medical AI devices are evaluated: limitations and recommendations from an analysis of FDA approvals. *Nat Med*, 27:582–584, 2021.

[275] Zacks Equity Research. Homepage. www.zacks.com/research/equity-research.php.

Index

Printed in the United States
by Baker & Taylor Publisher Services